Lecture Notes in Computer Science 12617

More information about this subseries at http://www.springer.com/series/7410

Xianfeng Zhao · Yun-Qing Shi ·
Alessandro Piva · Hyoung Joong Kim (Eds.)

Digital Forensics
and Watermarking

19th International Workshop, IWDW 2020
Melbourne, VIC, Australia, November 25–27, 2020
Revised Selected Papers

 Springer

Editors
Xianfeng Zhao 🆔
Institute of Information Engineering
Chinese Academy of Sciences
Beijing, China

Alessandro Piva 🆔
Department of Information Engineering
University of Florence
Florence, Italy

Yun-Qing Shi
Department of Electrical
and Computer Engineering
New Jersey Institute of Technology
Newark, NJ, USA

Hyoung Joong Kim 🆔
School of Cybersecurity
Korea University
Seoul, Korea (Republic of)

ISSN 0302-9743   ISSN 1611-3349   (electronic)
Lecture Notes in Computer Science
ISBN 978-3-030-69448-7   ISBN 978-3-030-69449-4   (eBook)
https://doi.org/10.1007/978-3-030-69449-4

LNCS Sublibrary: SL4 – Security and Cryptology

This Springer imprint is published by the registered company Springer Nature Switzerland AG
The registered company address is: Gewerbestrasse 11, 6330 Cham, Switzerland

Preface

The International Workshop on Digital-forensics and Watermarking (IWDW) is a premier forum for researchers and practitioners working on novel research, development and application of digital watermarking and forensics techniques for multimedia security. The 19th International Workshop on Digital-forensics and Watermarking (IWDW 2020) was organized by the Digital Research & Innovation Capability Platform at Swinburne University of Technology, Australia and the State Key Laboratory of Information Security at the Institute of Information Engineering, Chinese Academy of Sciences. It was held in Melbourne, Australia, during November 25–27, 2020. Although IWDW 2020 was held with the aid of an online conference system due to the persistent Covid-19 epidemic situation, it was very successful and as significant as the previous editions. More than 160 people attended the workshop online.

IWDW 2020 aimed at promoting research and development in both new and traditional areas of multimedia security. The organizers updated the topics of interest in the Call For Papers to reflect the new directions of emerging technologies such as AI-generated multimedia and detection of them, DeepFake videos and detection of them, convolutional neural networks and deep learning for multimedia security, and social media steganography. IWDW 2020 received 43 valid submissions. The decisions of the Technical Program Committee were made on a highly competitive basis. Only 20 submissions were accepted. The accepted papers cover many important topics in current research in multimedia security, and the presentations were organized into four sessions including "Steganography and Steganalysis", "Watermarking", "Multimedia Forensics" and "Security of AI-based Multimedia Applications". In addition, 3 invited keynotes including "Using the sensor noise model to design better steganographic schemes" by Dr. Patrick Bas, "DeepFake detection" by Dr. Wenbo Zhou, and "On the sharing-based model of steganography" by Prof. Xianfeng Zhao reported new advances.

We would like to thank all of the authors, committee members, reviewers, keynote speakers, volunteers and attendees. It is the participation of them all that made a wonderful and special IWDW again. And we appreciate the generous support from the organizers and sponsors. Finally, we hope that the readers will enjoy this volume and find it rewarding in providing inspirations and possibilities for future work.

December 2020

Xianfeng Zhao
Yun-Qing Shi
Alessandro Piva
Hyoung Joong Kim

Organization

General Chairs

Yun-Qing Shi	New Jersey Institute of Technology, USA
Yang Xiang	Swinburne University of Technology, Australia

Technical Program Chairs

Xianfeng Zhao	Chinese Academy of Sciences, China
Hyoung Joong Kim	Korea University, Korea
Alessandro Piva	University of Florence, Italy

Program Committee

Pradeep K. Atrey	University at Albany, USA
Mauro Barni	University of Siena, Italy
Patrick Bas	University of Lille, France
Marc Chaumont	University of Montpellier, France
Pedro Comesaña Alfaro	University of Vigo, Spain
Isao Echizen	National Institute of Informatics, Japan
Teddy Furon	Inria, France
Anthony T. S. Ho	University of Surrey, UK
Jiwu Huang	Shenzhen University, China
Yongfeng Huang	Tsinghua University, China
Xinghao Jiang	Shanghai Jiao Tong University, China
Xiangui Kang	Sun Yat-sen University, China
Sang-ug Kang	Sangmyung University, Korea
Stefan Katzenbeisser	University of Passau, Germany
Suah Kim	Korea University, Korea
Christian Kraetzer	Otto-von-Guericke University of Magdeburg, Germany
Minoru Kuribayashi	Okayama University, Japan
Bin Li	Shenzhen University, China
Xiaolong Li	Beijing Jiaotong University, China
Feng Liu	Chinese Academy of Sciences, China
Wojciech Mazurczyk	Warsaw University of Technology, Poland
Jiangqun Ni	Sun Yat-sen University, China
Akira Nishimura	Tokyo University of Information Sciences, Japan
Fernando Pérez-González	University of Vigo, Spain

William Puech	University of Montpellier, France
Pascal Schöttle	Management Center Innsbruck, Austria
Yun-Qing Shi	New Jersey Institute of Technology, USA
Luisa Verdoliva	University of Naples Federico II, Italy
Hongxia Wang	Sichuan University, China
Kai Wang	Grenoble-INP GIPSA-Lab, France
Andreas Westfeld	University of Applied Sciences Dresden, Germany
Yang Xiang	Swinburne University of Technology, Australia
Wei Qi Yan	Auckland University of Technology, New Zealand
Dengpan Ye	Wuhan University, China
Xinpeng Zhang	Fudan University, China
Weiming Zhang	University of Science and Technology of China, China
Yao Zhao	Beijing Jiaotong University, China
Guopu Zhu	Chinese Academy of Sciences, China

Organization Chairs

| Yang Xiang | Swinburne University of Technology, Australia |
| Hong Zhang | Chinese Academy of Sciences, China |

Organization Committee

Abby Xu	Insightek, Australia
Xiaolei He	Chinese Academy of Sciences, China
Yanfei Tong	Chinese Academy of Sciences, China

Additional Reviewers

Qi Chang
Rong Huang
Jian Li
Bo Ou
David Vázquez-Padín
Kai Zeng

Sponsors

 Chinese Academy of Sciences

 Swinburne University of Technology

 New Jersey Institute of Technology

 Springer

Contents

Steganography and Steganalysis

Multi-modal Steganography Based on Semantic Relevancy

Yuting Hu[1,2], Zhongliang Yang[1,2(✉)], Han Cao[1], and Yongfeng Huang[1,2]

[1] Department of Electronic Engineering,
Tsinghua University, Beijing 100084, China
huyt16@mails.tsinghua.edu.cn,
{yangzl15,caoh16}@tsinghua.org.cn, yfhuang@mail.tsinghua.edu.cn
[2] Beijing National Research Center for Information Science and Technology,
Beijing 100084, China

Abstract. Traditional steganography embeds confidential information by modifying the carrier at the symbol level, *e.g.*, the pixels of an image or the words of a text. Since modification traces will inevitably be left on the carrier, it is hard to resist the detection of the steganalysis algorithms. To address this problem, this paper proposes a novel steganographic framework called multi-modal steganography, which hides secret messages at the semantic level. In this framework, multi-modal covers are projected into a common semantic space, in which their relevancy can be measured. The confidential information can be embedded in the semantic relevancy among the covers with a relevancy-message mapping algorithm. By choosing and sending a series of original multi-modal covers, the secret messages are transmitted to the receiver. In this paper, we adopt text and image as the two modalities. A visual semantic embedding model is utilized to measure the relevancy between the texts and images. Both the theoretical analysis and experiments demonstrate that the proposed multi-modal steganography has good resistance to the existing steganalysis methods and high quality of concealment.

Keywords: Multi-modal steganography · Semantic relevancy · Visual semantic embedding · Relevancy-message mapping

1 Introduction

Steganography is the art and science of hiding confidential information within digital carriers [1,7,18]. There are three main types of steganography technologies according to [3], *i.e.*, steganography by cover modification, steganography by cover synthesis and steganography by cover selection.

Currently, steganography by cover modification is the most widely-used steganography technology [9,11,20]. Secret messages are embedded in the cover

This research is supported by the National Key R&D Program (2018YFB0804103) and the National Natural Science Foundation of China (No. U1705261 and No. U1836204).

X. Zhao et al. (Eds.): IWDW 2020, LNCS 12617, pp. 3–14, 2021.
https://doi.org/10.1007/978-3-030-69449-4_1

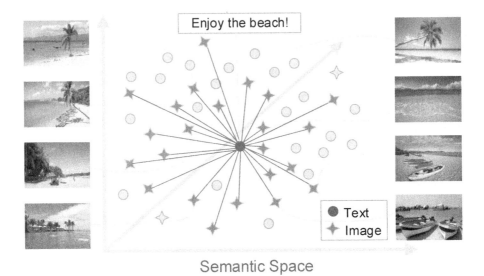

Fig. 1. An example of the distribution of two modalities (*i.e.*, images and texts) in a common semantic space.

by modifying the content of the cover such as the pixels of an image or the words of a text. The cover after modification is different from the original cover more or less. Thus, steganalysis methods are able to detect the existence of the secret messages based on the modification traces [4,13].

With the rapid development of deep learning, it is possible to generate texts and images automatically based on recurrent neural networks and generative adversarial networks. Steganography by cover synthesis has attracted more and more research interests [14,18]. However, the difference between the distributions of the generated covers and the original covers makes the steganalysis possible [16,17].

Steganography by cover selection, also named as coverless information hiding, embeds secret messages in a series of original covers without modification. A series of covers which contain the secret messages are chosen from a constructed database and sent to the receiver. Hash functions are designed to convert a cover into a binary sequence based on the local features of the cover such as intensity value [23], HOG [24] and SIFT [21]. Other method uses partial-duplicate image retrieval to transmit secret color image [22]. But semantics of the covers are not considered in these methods. The transmitted series of covers may be content-independent while the contents in a post on the social networks are likely to be relevant. Since these steganographic methods ignore the behavioral security [15], they may be detected by side channel steganalysis [8]. In order to control the topic of the images in each transmission, a coverless image steganography algorithm is proposed based on latent dirichlet allocation (LDA) topic classification

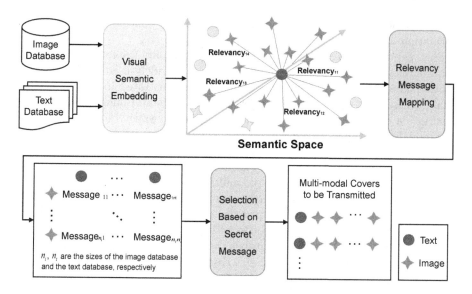

Fig. 2. The overall framework of the proposed multi-modal steganography.

[19]. However, the feature sequence of an image is fixed since it is constructed by the relation of discrete cosine (DC) coefficients between the adjacent blocks of the image. It is more secure and flexible if an image can represent various sequences in different transmissions.

Although the aforementioned methods employ various steganography technologies, all of them have a common point. That is they implement the steganography at the symbol level. These methods focus more on the low-level features such as pixels, intensity value, HOG, SIFT, DC coefficients of an image. Consequently, it is possible for the existing steganalysis tools to detect them somehow. Therefore, it is time to conduct steganography at the semantic level.

As we know, digital multimedia is popular in network transmission. Generally speaking, the modalities appearing on the same web page usually have higher relevancy in semantics than those appearing on the different occasion. Figure 1 presents an example of the distribution of two modalities (*i.e.*, images and texts) in a common semantic space. The distance between the relevant modalities is closer than that between the irrelevant ones. Meanwhile, there exist multiple pieces of relevant data of one modality such as image for the data of another modality such as text. This inspires us to hide the confidential information by utilizing the semantic relevancy. For a simple example, if a text have 16 relevant images, then transmitting the most relevant image can represent the secret messages $\{0,0,0,0\}$ while transmitting the least relevant image can represent $\{1,1,1,1\}$. Since the transmitted text and images are original and relevant in semantics, this kind of steganography can escape from the detection of the existing steganalysis methods. Two key difficulties are how to measure the semantic relevancy among the multiple modalities and how to conceal the confidential information in the semantic relevancy.

This paper proposes a multi-modal steganography framework based on semantic relevancy. Texts and images, the most two widely-used modalities, are adopted. The overall steganography framework is illustrated in Fig. 2. With a visual semantic embedding (VSE) algorithm [2], texts and images can be projected into a common semantic space, in which the relevancy between two arbitrary modality data can be measured. The semantic relevancy can be converted into a binary sequence with a relevancy-message mapping algorithm. After selecting the texts and images whose binary sequences are the same as the secret message segments, the sender transmits the multi-modal covers to the receiver. This work can be regarded as an extension version of the previous work of our research group [5]. In the previous work, we propose a basic multi-modal steganography (MM-Stega) framework. But the process of information extraction requires the receiver to share a common image database with the sender, which is a huge synchronization overhead. In this paper, this cost can be avoided with the proposed relevancy-message mapping algorithm.

The main contributions of this paper are concluded as follows. We propose a multi-modal steganography framework, which conduct steganography at the level of semantics instead of symbol. Rather than modify or generate a cover, we conceal the confidential information in the semantic relevancy among the original covers. Therefore, this kind of steganography can escape from detection of the existing steganalysis methods. Moreover, it will arouse lower suspicion to transmit covers which are relevant in semantics since it obeys the routine of the multimedia on the network transmission. In addition, the sequence represented by an image can be various when different texts are chosen, which brings more flexibility and security.

The rest of this paper is organized as follows. The proposed method is described in detail in Sect. 2. The experimental results and analysis are shown in Sect. 3. Finally, Sect. 4 concludes the paper.

2   The Proposed Multi-modal Steganography

In this section, we will illustrate the proposed multi-modal steganography framework. The most two widely-used modalities, texts and images, are adopted. The flow chart of the multi-modal steganography for covert communication is represented in Fig. 3. Before the covert communication, it is necessary to set up a text database and an image database. The two databases are constructed by selecting a number of texts and images from the Twitter100k dataset [6], which is a large-scale text-image dataset. A visual semantic embedding model [2] trained on the Twitter100k dataset [6] is utilized to measure the relevancy between a text and an image. We propose a relevancy-message mapping algorithm which can map an image to a binary sequences for a given text.

The sender first converts the secret messages into a bit stream and divides it into n binary segments with the equal length k, where n and k are set according to the amount of the secret message. k is acknowledged by both sender

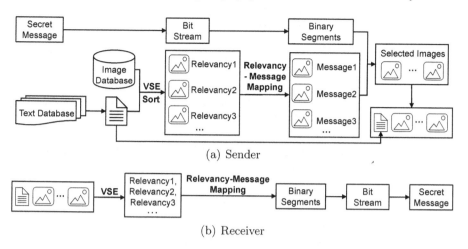

(a) Sender

(b) Receiver

Fig. 3. The flow chart of the proposed multi-modal steganography for covert communication at (a) sender and (b) receiver.

and receiver of the covert communication. After choosing a text randomly from the text database, the relevancies between the text with all the images in the image database are measured with the visual semantic embedding model. All the images are sorted on the basis of the relevancies. Afterwards, each image represents a k-bits binary sequence according to the relevancy-message mapping algorithm. Then, images whose binary sequences are the same as the secret message segments are selected. The images which have higher relevancies with the selected text are prior to being selected and each image can only be selected at most once. Finally, the combination of the text and the images is conveyed to the receiver in order.

The receiver shares the same visual semantic embedding model with the sender. After receiving a combination of one text and n images, the relevancies between the images and the text are measured with the visual semantic embedding model. Then, the binary sequence of each image is computed with the relevancy-message mapping algorithm. After splicing the binary sequences together into a bit stream in the order of the received images, the confidential information is obtained by converting the bit stream.

To sum up, the main parts of the proposed method are the visual semantic embedding model, relevancy-message mapping algorithm, information hiding algorithm and information extraction algorithm.

2.1   The Visual Semantic Embedding Model

Visual semantic embedding model [2] is utilized to measure the semantic relevancy between the two modalities. As illustrated in Fig. 4, the visual semantic embedding model is composed of a text-embedding network and an image-embedding network. The two embedding networks project the text and image

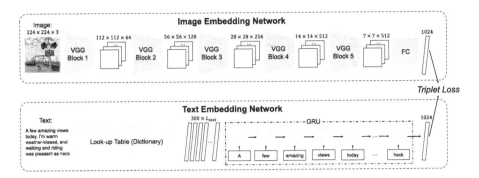

Fig. 4. The overall network structure of the visual semantic embedding model, which consists of a text-embedding network and an image-embedding network.

into a common semantic space, where the embeddings of the two modalities which are relevant in semantics are close to each other. The details of the visual semantic embedding model are the same as [5].

2.2 Relevancy-Message Mapping Algorithm

The relevancy between the two modalities is a decimal ranging from -1 to 1. We denote the digits after decimal point as $\{d_1, d_2, ..., d_n, ...\}$. Then the digits $\{d_1, d_2, ..., d_n, ...\}$ can be converted to a binary sequence $\{b_1, b_2, ..., b_n, ...\}$ based on the parity of each digit. That is,

$$b_n = \begin{cases} 0, & if \ d_n \ is \ even \\ 1, & if \ d_n \ is \ odd \end{cases} \tag{1}$$

The message represented by an image for a given text is $\{b_i, ..., b_{i+k-1}\}$, where $i, k \in \{1, 2, 3, ...\}$. i will be set in the light of the distribution of the binary sequence $\{b_1, b_2, ..., b_n, ...\}$. k is the length of the message and is shared by both sender and receiver.

In order to know the distribution of the binary sequence $\{b_1, b_2, ..., b_n, ...\}$, we calculate the entropy and the joint entropy of the parity of each digit on the dataset described in Sect. 3.2 according to the following formulas.

$$H(B_n) = - \sum_{b_n \in \{0,1\}} p(b_n) log p(b_n) \tag{2}$$

$$H(B_i, ..., B_{i+k-1}) = - \sum_{b_i, ..., b_{i+k-1} \in \{0,1\}^k} p(b_i, ..., b_{i+k-1}) log p(b_i, ..., b_{i+k-1}) \tag{3}$$

The results of the entropy $H(B_i)$ and the joint entropy $H(B_i, ..., B_{i+k-1})$ are presented in Table 1 and Table 2, respectively. As we can see,

$$|H(B_i) - 1| < \epsilon, \ when \ i \geq 3, \ \epsilon = 0.01 \tag{4}$$

$$|H(B_i, ..., B_{i+k-1}) - \sum_{n=i}^{i+k-1} H(B_n)| < \epsilon, \; when \; i \geq 3, \; \epsilon = 0.02 \qquad (5)$$

It means that the parity of the third digit and the latter digits of the relevancy is nearly random and independent, which can provide high diversity for the message represented by the image. In this paper, the message represented by an image for a given text is $\{b_4, ..., b_{4+k-1}\}$, where k is adjustable.

Table 1. The results of the entropy $H(B_i)$.

i	1	2	3	4	5	6	7	8	9
$H(B_i)$	0.36	0.57	1.00	1.00	1.00	1.00	1.00	1.00	1.00

Table 2. The results of the joint entropy $H(B_i, ..., B_{i+k-1})$.

i	1	2	3	4	5
$H(B_i, B_{i+1})$	1.70	1.57	2.00	2.00	2.00
$H(B_i, B_{i+1}, B_{i+2})$	2.69	2.60	3.00	3.00	3.00
$H(B_i, B_{i+1}, B_{i+2}, B_{i+3})$	3.69	3.63	3.99	3.99	3.99
$H(B_i, B_{i+1}, B_{i+2}, B_{i+3}, B_{i+4})$	4.67	4.68	4.98	4.98	4.98

2.3 Information Hiding Algorithm

There are five steps in the information hiding procedure, which will be introduced in detail as follows.

Confidential Messages Preprocessing. We convert the confidential information into a binary sequence. The binary sequence is divided into n segments with the equal length k. n is the number of the attached images to the selected text. k is the number of bits which are represented by an image. n and k are set according to the amount of the confidential information, the size of the image database and the limitations of the data transmission platform.

Text Selection. We choose a text randomly from the text database.

Relevancy Measurement. For each image in the image database, we measure the relevancy between the image with the selected text based on the well-trained visual semantic embedding model. All the images in the image database are sorted according to their relevancies to the selected text.

Relevancy-Message Mapping. The message represented by each image for the selected text is calculated with the relevancy-message mapping algorithm.

Image Selection. We compare the confidential information segments with the messages represented by the images in the descend order of the relevancy between the images and the selected text. The images which represent the same messages as the confidential information segments are selected. The image having higher semantic relevancy with the selected text is prior to be picked out.

After these five steps, we will get a text and n images which hide the confidential information. Then these original multi-modal covers are transmitted to the receiver.

2.4   Information Extraction Algorithm

The receiver shares the visual semantic embedding model and the parameter k with the sender. After receiving the text and n images, three steps are operated to extract the confidential information.

Relevancy Measurement. The relevancies between the received images and the text are calculated with the shared visual semantic embedding model.

Relevancy-Message Mapping. The relevancy between the received image and the text is converted to the message represented by the image according to the relevancy-message mapping algorithm.

Confidential Information Recovery. All the messages represented by the received images are spliced together into a bit stream. Finally, the bit stream is transformed to the confidential information.

3   Experiments and Analysis

The performance of the proposed multi-modal steganography can be evaluated from four aspects, *i.e.*, hiding capacity, semantic relevancy, resistance to the steganalysis methods and complexity.

3.1   Hiding Capacity

The hiding capacity of the proposed steganography is proportional to the number of companied images with a text and the bits contained by each image. Assume each text is companied with n images and each image represents k-bit binary sequence, then the hiding capacity of the proposed method is $k \times n$ bits. The hiding capacity of the proposed method is adjustable. There is no obvious limitation of n. Though k is limited by the size of the image database, the image database can be enriched with the public Twitter100k dataset without much effort.

3.2   Semantic Relevancy

The imperceptibility of the proposed method depends on the semantic relevancy between the selected two modalities. The more relevant between the text and the images, the less suspicion will be aroused. We conduct the experiment on the dataset proposed in [5], which contain 100 texts and 1,000 images. All the texts and images are labeled with several tags such as people, clothes and flower. The relevancy between a text and an image can be evaluated by matching the tags. If a text and an image contain a common tag, the image is regarded as relevant to the text in the experiment. For a given text, we define the relevant rate as the ratio of the number of the relevant companied images to the number of all the companied images with the text.

In our experiment, binary sequence is used as the confidential information. For a given length of confidential information, all the possible binary sequences are tested in the experiment. Since the number of the possible binary sequences increases exponentially with n and k, we adopt $n = 1, 2, 3, 4$ and $k = 1, 2, 3, 4, 5$. The relevant rate is calculated for each text in the text database and the result of the average relevant rate is presented in Table 3.

Table 3. The average relevant rates at different values of k and n.

n	k				
	1	2	3	4	5
1	0.87	0.85	0.82	0.80	0.78
2	0.85	0.84	0.82	0.80	0.78
3	0.85	0.84	0.81	0.80	0.78
4	0.84	0.83	0.81	0.80	0.78

As shown in Table 3, the average relevant rate is greater than 0.78 when $k \leq 5$ and $n \leq 4$, which achieves a relative satisfying relevancy. Moreover, the average relevant rate decreases very slightly (smaller than 0.01) with the increase of n. Therefore, we can use more images to convey more confidential information without much sacrifice of the imperceptibility.

Some examples of the proposed steganography are given in Fig. 5. Each secret message is an English word of five characters, which can be transformed into 40-bit binary sequence. The values of n and k are set to 8 and 5, respectively. That is, each text is companied with 8 images and each image represents 5 bits. It can be seen that the companied images are relevant to the text and the combination of the two modalities is very natural.

3.3   Resistance to the Steganalysis Methods

Ideal steganography should have good resistance to the detection of the steganalysis methods. However, existing steganalysis methods can successfully detect the

Secret Message: *China*
It's always good to see your friends from sLOVEenija in Miami !!! #welcomeToMiami

Secret Message : *Japan*
Gorgeous scenery. Beautiful weather. FREEDOM!

Secret Message : *Korea*
Today began with a colourful start #toronto #sunrise

Fig. 5. Examples of the proposed multi-modal steganography. Each text is companied with 8 images and each image represents 5 bits.

steganography by modification [10, 12] and steganography by synthesis [16, 17]. In contrast to the previous steganography which performs information hiding at the symbol level, the proposed multi-modal steganography embeds confidential information based on the relevancy between two modalities at the semantic level. All the covers keep natural and original without modification or synthesis. In addition, the selected covers are highly relevant in semantics. Therefore, the proposed multi-modal steganography is unlikely to be detected by the existing steganalysis methods.

3.4   Complexity

Compared with the previous MM-Stega which embeds the secret messages in the relative values of the semantic relevancies [5], the method proposed in this work utilizes the value of the semantic relevance directly. Therefore, the space complexity is reduced to a large extent because there is no need for the receiver to have an image database same as the sender in the proposed method. Moreover, in the previous MM-Stega, all the semantic relevancies between the received text and each image in the image database are required to be computed. This computational complexity is decreased in our method since only the relevancies between the received text and the received images are needed at the receiver end.

4   Conclusion

This paper puts forward a novel multi-modal steganography framework based on semantic relevancy. A visual semantic embedding model is adopted to measure the semantic relevancy between the two modalities, *i.e.*, texts and images. A relevancy-message mapping algorithm is proposed to calculate the messages represented by an image for a given text. The sequence represented by an image can be various given different texts. Since the confidential information is hidden in the relevancy between the two modalities at the semantic level, there is no need to modify or generate the covers. Therefore, the proposed method can effectively resist the detection of the existing steganalysis methods. Experiments verify that the selected covers are highly relevant in semantics, which guarantees good security. Moreover, the space complexity and the computational complexity of the proposed method are much reduced compared with the previous MM-Stega since there is no need for receiver to keep a same image database with the sender.

References

1. Du, Y., Yin, Z., Zhang, X.: Improved lossless data hiding for JPEG images based on histogram modification. Comput. Mater. Continua **55**(3), 495–507 (2018)
2. Faghri, F., Fleet, D.J., Kiros, J.R., Fidler, S.: VSE++: improving visual-semantic embeddings with hard negatives. arXiv preprint arXiv:1707.05612 (2017)
3. Fridrich, J.: Steganography in Digital Media: Principles, Algorithms, and Applications. Cambridge University Press, Cambridge (2009)
4. Fridrich, J., Kodovsky, J.: Rich models for steganalysis of digital images. IEEE Trans. Inf. Forensics Secur. **7**(3), 868–882 (2012)
5. Hu, Y., Li, H., Song, J., Huang, Y.: MM-Stega: multi-modal steganography based on text-image matching. In: Sun, X., Wang, J., Bertino, E. (eds.) ICAIS 2020. CCIS, vol. 1254, pp. 313–325. Springer, Singapore (2020). https://doi.org/10.1007/978-981-15-8101-4_29
6. Hu, Y., Zheng, L., Yang, Y., Huang, Y.: Twitter100k: a real-world dataset for weakly supervised cross-media retrieval. IEEE Trans. Multimed. **20**(4), 927–938 (2018)
7. Huang, Y., Liu, C., Tang, S., Bai, S.: Steganography integration into a low-bit rate speech codec. IEEE Trans. Inf. Forensics Secur. **7**(6), 1865–1875 (2012)
8. Li, L., Zhang, W., Chen, K., Zha, H., Yu, N.: Side channel steganalysis: when behavior is considered in steganographer detection. Multimed. Tools Appl. **78**(7), 8041–8055 (2019)
9. Li, X., Yang, B., Cheng, D., Zeng, T.: A generalization of LSB matching. IEEE Sig. Process. Lett. **16**(2), 69–72 (2009)
10. Lie, W.N., Lin, G.S.: A feature-based classification technique for blind image steganalysis. IEEE Trans. Multimed. **7**(6), 1007–1020 (2005)
11. Westfeld, A.: F5—A steganographic algorithm. In: Moskowitz, I.S. (ed.) IH 2001. LNCS, vol. 2137, pp. 289–302. Springer, Heidelberg (2001). https://doi.org/10.1007/3-540-45496-9_21
12. Wu, S., Zhong, S., Liu, Y.: Deep residual learning for image steganalysis. Multimed. Tools Appl. **77**(9), 10437–10453 (2018)
13. Xu, G., Wu, H., Shi, Y.: Structural design of convolutional neural networks for steganalysis. IEEE Sig. Process. Lett. **23**(5), 708–712 (2016)
14. Yang, Z., Guo, X., Chen, Z., Huang, Y., Zhang, Y.: RNN-Stega: linguistic steganography based on recurrent neural networks. IEEE Trans. Inf. Forensics Secur. **14**(5), 1280–1295 (2018)

15. Yang, Z., Hu, Y., Huang, Y., Zhang, Y.: Behavioral security in covert communication systems. arXiv preprint arXiv:1910.09759 (2019)
16. Yang, Z., Huang, Y., Zhang, Y.: A fast and efficient text steganalysis method. IEEE Sig. Process. Lett. **26**(4), 627–631 (2019)
17. Yang, Z., Wang, K., Li, J., Huang, Y., Zhang, Y.: TS-RNN: text steganalysis based on recurrent neural networks. IEEE Sig. Process. Lett. **26**(12), 1743–1747 (2019). https://doi.org/10.1109/LSP.2019.2920452
18. Yang, Z., Zhang, S., Hu, Y., Hu, Z., Huang, Y.: VAE-Stega: linguistic steganography based on variational auto-encoder. IEEE Trans. Inf. Forensics Secur. **16**, 880–895 (2020)
19. Zhang, X., Peng, F., Long, M.: Robust coverless image steganography based on DCT and LDA topic classification. IEEE Trans. Multimed. **20**(12), 3223–3238 (2018)
20. Zhang, Y., Ye, D., Gan, J., Li, Z., Cheng, Q.: An image steganography algorithm based on quantization index modulation resisting scaling attacks and statistical detection. Comput. Mater. Continua **56**(1), 151–167 (2018)
21. Zheng, S., Wang, L., Ling, B., Hu, D.: Coverless information hiding based on robust image hashing. In: Huang, D.-S., Hussain, A., Han, K., Gromiha, M.M. (eds.) ICIC 2017. LNCS (LNAI), vol. 10363, pp. 536–547. Springer, Cham (2017). https://doi.org/10.1007/978-3-319-63315-2_47
22. Zhou, Z., Mu, Y., Wu, Q.J.: Coverless image steganography using partial-duplicate image retrieval. Soft Comput. **23**(13), 4927–4938 (2019). https://doi.org/10.1007/s00500-018-3151-8
23. Zhou, Z., Sun, H., Harit, R., Chen, X., Sun, X.: Coverless image steganography without embedding. In: Huang, Z., Sun, X., Luo, J., Wang, J. (eds.) ICCCS 2015. LNCS, vol. 9483, pp. 123–132. Springer, Cham (2015). https://doi.org/10.1007/978-3-319-27051-7_11
24. Zhou, Z., Wu, Q.J., Yang, C.N., Sun, X., Pan, Z.: Coverless image steganography using histograms of oriented gradients-based hashing algorithm. J. Internet Technol. **18**(5), 1177–1184 (2017)

Variable Rate Syndrome-Trellis Codes for Steganography on Bursty Channels

Bingwen Feng[1,2(✉)], Zhiquan Liu[1,3], Kaimin Wei[1], Wei Lu[4], and Yuchun Lin[5]

[1] College of Information Science and Technology,
Jinan University, Guangzhou 510632, China
bingwfeng@gmail.com, {zqliu,kmwei}@jnu.edu.cn
[2] State Key Laboratory of Information Security, Institute of Information Engineering,
Chinese Academy of Sciences, Beijing 100093, China
[3] Guangdong Key Laboratory of Intelligent Information Processing
and Shenzhen Key Laboratory of Media Security, Shenzhen 518060, China
[4] School of Computer Science and Engineering,
Guangdong Province Key Laboratory of Information Security Technology,
Ministry of Education Key Laboratory of Machine Intelligence and Advanced
Computing, Sun Yat-sen University, Guangzhou 510006, China
luwei3@mail.sysu.edu.cn
[5] Guangzhou Institute of Science and Technology,
Guangzhou 510006, China
ychLin@gmail.com

Abstract. This paper presents a type of variable rate syndrome-trellis codes (VR-STC) for bursty channels. It can embed message bits with two different embedding rates. In the embedding, a cover vector is sliced into segments, and the embedding rates for each segment depend on the local channel distribution. The parities of stego segments are exploited to indicate the selected embedding rates according to a mapping function between parities and embedding rates. The core of the VR-STC is the parity-aware encoder, which can simultaneously output two candidate stego segments with different embedding rates and opposite parities, either of which can be used to constitute the final stego vector. A Viterbi algorithm is also suggested to find the closed stego segments. Besides, the mapping function between parities and embedding rates is designed by minimizing the embedding costs on a down sampled version of the cover vector. It can further improve the undetectability of the VR-STC. Experimental results on artificial signals and binary images suggest that the proposed VR-STC can provide high success rate of embedding and reduce the embedding cost on bursty channels.

Keywords: Variable rate syndrome-trellis codes (VR-STC) ·
Steganography · Bursty channel · Dynamical embedding rate ·
Parity-aware encoder

© Springer Nature Switzerland AG 2021
X. Zhao et al. (Eds.): IWDW 2020, LNCS 12617, pp. 15–30, 2021.
https://doi.org/10.1007/978-3-030-69449-4_2

1   Introduction

Steganography is a technique about covert communication. One of its important properties is undetectability, which calls for that the warden should not be able to detect the existence of the embedded messages [1]. Besides circumventing detection from the warden, it is also wished to maximize the embedding capacity to face practical communication requirement. For this reason, undetectability and capacity should be carefully balanced in designing steganographic schemes.

Modern steganographic schemes usually formulate the balance of steganography as a minimization problem of a heuristically chosen distortion measurement. Since the undetectability experimentally shows a strong relationship with the media content, adaptive steganography takes into account the difference among image region. In [2], web paper codes are suggested to avoid modifying "wet" pixels, i.e., pixels not suitable for carrying message bits. Syndrome-trellis codes (STCs) proposed in [3] provide a general methodology for embedding while minimizing an arbitrary additive distortion function near the theoretical bound. Steganographic polar codes (SPCs) proposed in [4] suggest another near optimal steganographic coding with low embedding complexity. Benefitting from these proposals, adaptive steganographic schemes can focus on designing cost functions [5,6].

These steganographic schemes usually assume that the cover follows a stationary distribution [1], which can be then used to select embedding parameters [2–4], or to define cost functions [5,6]. In the case of STC framework, message bits are embedded with an embedding rate fixed by the employed parity check matrix, and the embedding changes follow a multivariate Bernoulli distribution [7]. However, multimodal signals in the real world sometimes have strong time/space dependent behavior. For example, semantically meaningless frames [8] or static scenes [9] may crop up in a video sequence, and the voice may suddenly change from talk-spurt to silent-period in VoIP transmission [10]. These channels are not always suitable for steganography, and sometimes usable host signals appear unpredictably. This paper models them as bursty channels [11]. They assumes that the channel would be bad for a long period of time once it becomes bad. Take the binary image shown in Fig. 1 as an example. It can be observed a large contiguous area of flatten region. Intuitively, image block (a) could capture more message bits than (b), and block (c) only contains wet pixels and should not be used. We could query the random permutation to overcome the changing local distributions in this image. However, it requires capturing a large number of cover pixels for a cover vector so that the distributions among cover vectors vary little, which is not suitable for real time applications. Furthermore, the embedding rate has to be low enough to fit the worst embedding case, e.g., a cover vector containing block (c) in Fig. 1, which wastes the embedding capacity.

Regarding to the error correction code (ECC) applications, the changing local distributions presented in bursty channels can be overcome by protecting data with different importance [12–15]. In [14], the parity check bits on sub-blocks of orthogonal Latin square codes are used to protect part of the word with

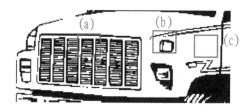

Fig. 1. Demonstration of the bursty characteristic in a binary image.

double error correction and the other part with single error correction and double error detection. In [13], a two-level burst error correcting unequal protection code is proposed based on an existing construction of burst error correcting and single-bit error correcting codes. It can protect different parts of the word from different level burst error. However, the protection level is unchangeable for each part of a word in these schemes. In [12], with a selection signal, the data length of ECC can be dynamically changed to focus on the relatively more important bit parts when the number of failures exceeds the error correction capability. In [15], the reconfigurable ECC can adaptively change error correction capability depending on the severities of static, spatial and temporal variations. They motivate us to make the embedding rate of a steganographic scheme dynamically adapt to the varying local signal distribution. Similar to the above schemes, a selection signal indicating the embedding mode is in need. However, we can not embed these signal bits directly, because they present regular statistics easy to be detected. Moreover, inserting them into the message sequence would reduce the embedding capacity seriously. In view of these, we present a construction of adaptive steganographic codes, where embedding the selection signal affects the embedding efficiency marginally.

In this paper, a type of variable rate syndrome-trellis codes (VR-STC) is proposed for the bursty channel. Their embedding rates can be dynamically adjusted to adapting to the local cover distribution. It is achieved by the proposed parity-aware encoder. This encoder can simultaneously generate two stego segments with different embedding rates and opposite parities acting as indicates of embedding rates. One can determine the embedding rate according to the local embedding cost, and select the stego segment with the corresponded parity. A Viterbi algorithm is suggested to realize the parity-aware encoder. Furthermore, the mapping between parities and embedding rates is designed by minimizing the embedding costs on a type of down sampled images. It can further improve the undetectability of the VR-STC. Experimental results on artificial signals and binary images suggest that the proposed VR-STC can provide high success rate of embedding and reduce the embedding cost on bursty channels.

2   Variable Rate STC

The variable rate STC (denoted as VR-STC) embeds message bits with two alternative embedding rates, ρ_h and ρ_l. Between them which one is selected just now is indicated by the parity of the generated stego segment. This is carried out by a parity-aware encoder, which divides a cover vector into segments, and output two candidate stego segments with different embedding rates and opposite parities for each segment. Herein the parity of a binary sequence means its even or odd quality. Its value is 1 if the total number of "1" in the sequence is odd. When the embedding cost of the stego segment obtained by the higher embedding rate is unacceptable, e.g., it modifies some "wet" pixels, the other one will be chosen as a part of the encoder output.

2.1   Parity-Aware Encoder

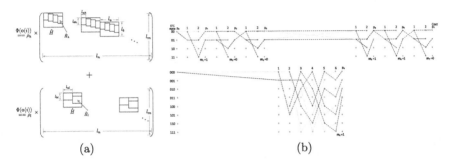

(a)   (b)

Fig. 2. a) is an example of the parity-check matrix H formed from two sub-submatrices $\hat{H}_h$ and $\hat{H}_l$. b) demonstrates the trellis constructed by the H shown in a), where a solid edge indicates the end of a path with even-parity, and a dash edge indicates odd-parity.

Instead of determining whether an embedding cost is acceptable, we force the parity of every stego segment to be a given value. This can provide flexible configuration, because it allows us to choose appropriate embedding costs, or to design secure mapping between parities and embedding rates.

The parity-aware encode is formed as a binary linear code. Its parity-check matrix H is reconfigurable to support changing embedding rates. It is achieved by adjusting the submatrices of H during runtime. The structure of H is designed as an extension of that in STC [3]. Given cover vector $\mathbf{x} \in \{0,1\}^{l_n \times 1}$ and message vector $\mathbf{m} \in \{0,1\}^{l_m \times 1}$, $H \in \{0,1\}^{l_n \times l_m}$ is constructed by placing a set of small submatrices, $\hat{H} \in \{0,1\}^{l_a \times 1_b}$, along its diagonal. Further, $\hat{H}$ is composed of smaller sub-submatrices $\hat{H}_l \in \{0,1\}^{l_{al} \times l_{bl}}$ or $\hat{H}_h \in \{0,1\}^{l_{ah} \times l_{bh}}$, where $l_{al} > l_{ah}$. Sub-submatrices with the same structure, whether adjacent to each other or not, are shifted down by one row, as shown in Fig. 2(a). It can be observed that the two sub-submatrices provide two embedding rates: $\rho_h = l_n/l_{ah}$ and $\rho_l = l_n/l_{al}$. Note that the widths of all the submatrices are same and set with a common multiple of l_{al} and l_{ah}.

Support that the two sub-submatrices have been pre-negotiated. Then the encoder is to find a stego vector $\mathbf{y} \in \{0,1\}^{l_n \times 1}$ by optimizing:

$$\mathbf{y} = \arg\min_{H\mathbf{y}'=\mathbf{m}} D(\mathbf{x}, \mathbf{y}') \tag{1}$$

$$s.t. \ \ H = \sum_i \mathrm{Cst}(i, \Phi(\mathbf{o}(i)), \hat{H}_h, \hat{H}_l) \tag{2}$$

$$H\mathbf{y} = \mathbf{m} \tag{3}$$

$$R\mathbf{y} = \mathbf{o} \tag{4}$$

where $\mathbf{o} \in \{0,1\}^{l_r \times 1}$, $l_r = l_n/l_a$, is a binary sequence storing the specified parities of each stego segment. Note that $\mathbf{o}$ can be arbitrarily adjusted during runtime, and is not required to transmit to the receiver. $\Phi()$ defines a mapping function which selects an embedding rate from $\{\rho_h, \rho_l\}$ according to the input parity. We will discuss its design in Sect. 3.2. $R \in \{0,1\}^{l_r \times l_n}$ is of form:

$$\begin{pmatrix} 1 \cdots 1 & & & 0 \\ & 1 \cdots 1 & & \\ & & \ddots & \\ 0 & & & 1 \cdots 1 \end{pmatrix}^T \tag{5}$$

where the number of "1" in each column is equal to l_a.

Function $\mathrm{Cst}(i, \Phi(\mathbf{o}(i)), \hat{H}_h, \hat{H}_l)$ in Eq. (1) is used to generate H by using submatrices $\hat{H}$, each of which is constructed by $\hat{H}_l$ if $\Phi(\mathbf{o}(i)) = \rho_l$, or by $\hat{H}_h$ if $\Phi(\mathbf{o}(i)) = \rho_h$. Figure 2(a) shows an example where $\Phi(\mathbf{o}(i)) = \rho_h, \rho_l, \rho_h, \rho_h, \rho_l, \cdots$ for $i = 1, 2, 3, 4, 5, \cdots$.

Using a structure of H similar to the STC' allows us to represent every solution of Eq. (1) as a path through the syndrome trellis of H. This trellis consists of two parallel graphs corresponded to ρ_h and ρ_l, respectively. The one corresponded to ρ_h consists of state grids of $(l_{ah} + 1)$ columns and $2^{l_{b,h}}$ rows. The other consists of grids of $(l_{al} + 1)$ columns and $2^{l_{bl}}$ rows.

We know that the STC associates each node with a cost value storing the minimal cost and a path index used to trace the previous path [3]. In the proposed encoder, each node has to store two incoming paths with opposite parities in order to find a path with a specified parity. As a result, the proposed encoder needs to double the storage space compared with the STC.

2.2 Description of Viterbi Algorithm

We use the Viterbi algorithm to find the closed stego through the trellis. It is similar to STC, except that the shortest path is now parity-aware. The algorithm consists of the forward and the backward parts.

The forward part is used to construct the trellis. Each node in the trellis stores two path indices with different parities and their cost. Supposing we are at the i-th node in the j-th column of the k-th block, we use $\mathbf{c}_{\mathrm{odd}}(i)$ and $\mathbf{p}_{\mathrm{odd}}(i)$

to denote the minimal cost and corresponded path index, respectively, for the incoming path with parity 1. Similarly, $\mathbf{c}_{\text{even}}(i)$ and $\mathbf{p}_{\text{even}}(i)$ are for the incoming path with parity 0. There are two edges leaving this node. One is labeled with 1, corresponding to a stego element with value 1, and the other is labeled with 0. Each of them updates not only both two paths' costs, but also their parities. For example, assume that there is an edge labeled with 1 that connects this node to the i'-th node in the $(j+1)$-th column, meanwhile the j-th cover element is 0 and assigned embedding cost w. Then this edge will increase the two paths' weight by w and flip their parities, because it alters the cover element and changes the number of "1" in the stego vector. As a result, the candidate cost values and path indices for the i'-th node in the $(j+1)$-th column provided by this edge are

$$\mathbf{c}'_{\text{odd}}(i') = \mathbf{c}_{\text{even}}(i) + w \tag{6}$$

$$\mathbf{p}'_{\text{odd}}(i') = i \times 2 \tag{7}$$

$$\mathbf{c}'_{\text{even}}(i') = \mathbf{c}_{\text{odd}}(i) + w \tag{8}$$

$$\mathbf{p}'_{\text{even}}(i') = i \times 2 + 1 \tag{9}$$

Note that we use the last bits of $\mathbf{p}'_{\text{odd}}(i')$ and $\mathbf{p}'_{\text{even}}(i')$ to indicate their prefixed paths' parities. For example, the last bit of $\mathbf{p}'_{\text{odd}}(i')$ in Eq. (7) means its prefixed path has even parity. There are 2 edges entering a node, each of which can provide 2 candidate paths with opposite parities. The one with a bigger weight will be removed if two paths have the same parity. If the closest two paths with different parities are provided by different edges, both of the entering edges will be preserved. Therefore, each node in the trellis has at most two incoming paths.

The cover is divided into segments of length l_a. Each time one segment is used to generate two parallel graphs of the trellis simultaneously. It can generate l_a/l_{al} grid blocks in the graph corresponded to $\hat{H}_l$, and l_a/l_{ah} blocks in the graph corresponded to $\hat{H}_h$. We assume the value of $\mathbf{o}(k)$, namely which graph is to be used, can be determined after the blocks in both graphs have been generated. This is reasonable, since the closest path through a block in a graph can be determined after the last column of this block has been generated. Thus we can judge whether its obtained stego segment is acceptable. Blocks not in use are removed from the graph. The designation of $\mathbf{o}$ is detailed in Sect. 3.2.

Figure 2(b) demonstrates an example of a constructed trellis, where the chosen submatrices are same as that in Fig. 2(a). In the figure there are two graphs. The above corresponds to sub-submatrix $\hat{H}_h$ and the below corresponds to $\hat{H}_l$. Grids in the same graph may be not adjacent. Further, some nodes may have entering edges from two previous nodes.

The backward part of the algorithm is to trace the closest paths through the two graphs. Which graph is to be used depends on the mapping function $\Phi()$ and specified parity vector $\mathbf{o}$ in Eq. (1). The path tracing starts from the reachable nodes in the last column of the last block in the graph specified by $\Phi(\mathbf{o}(l_r))$. A state is used to trace the path through this graph. If, for example, $\mathbf{o}(l_r) = 1$ and $\Phi(\mathbf{o}(l_r)) = \rho_h$, we choose the node with the minimum weight as the final state, saying $\hat{i}$, and use $\mathbf{p}_{\text{odd}}(\hat{i})$ associated with the last column of the l_r-th block generated by $\hat{H}_h$ to trace the previous nodes. Recall that the parity of the path traced next is indicated by the last bits of $\mathbf{p}_{odd}(\hat{i})$. Therefore, we should use $\mathbf{p}_{even}(\lfloor \mathbf{p}_{odd}(\hat{i})/2 \rfloor)$ associated with the last column but one to trace the earlier nodes if the last bit of $\mathbf{p}_{odd}(\hat{i})$ is 0.

There are two graphs in the trellis. Each graph has its own last block. Consequently, The backward part should use two states to trace two paths in different graphs. After l_a/l_{ah} blocks generated by $\hat{H}_h$ (or l_a/l_{al} blocks generated by $\hat{H}_l$) have been processed, $\Phi(\mathbf{o}(l_r - l_a/l_{al}))$ (or $\Phi(\mathbf{o}(l_r - l_a/l_{ah}))$) is used to select the graph to be processed next.

2.3   Corresponding Decoder

The corresponding decoder is simple. It divides the stego vector into segments of length l_a, and then calculates their parities. For the segment with parity 1, it uses the sub-submatrix, $\hat{H}_l$ or $\hat{H}_h$, corresponded to $\Phi(1)$ to extract message bits. One stego segment can extract l_a/l_{al} message bits if using $\hat{H}_l$, or l_a/l_{ah} message bits otherwise. For the sake of simplifying the message extraction procedure, all the stego segments generated by the same sub-submatrix can be combined together and input into the original STC extractor to get the message bits embedded.

3   Implementation Details

3.1   Submatrix Selection

The submatrix $\hat{H}$ should be well defined according to the considered cover images. Embedding message bits usually starts with assigning embedding cost $\mathbf{w}$ for the cover vector $\mathbf{x}$ with the aid of certain cost function. Those wet elements will be assigned with $\mathbf{w}(i) = \infty$ in this step. The proposed encoder then divides $\mathbf{x}$ into segments. Some of them present typical cover characteristics, i.e., they do not have too many cases of $\mathbf{w}(i) = 0$ or $\mathbf{w}(i) = \infty$ [16], while the other may contain a large number of wet elements.

As shown in Fig. 2(a), $\hat{H}$ can be constructed by either $\hat{H}_h$ or $\hat{H}_l$. Between them $\hat{H}_h$ provides a high embedding rate, and thus can be used in typical cover segments to ensure the embedding efficiency. Therefore, the selection of $\hat{H}_h$ follows the common suggestions in [3].

The sub-submatrix $\hat{H}_l$ is used for those very wet segments. We consider the extreme case where $dura$ in every $(dura + 1)$ cover elements are wet. In this case the width of $\hat{H}_l$ should satisfy $l_{al} \geq (dura + 1)$ to guarantee that the processed elements contains at least one dry element. Further, it can be observed that, in the Viterbi algorithm, the i-th message bit being embedded is associated with the current and previous $(\min\{l_{bl}, i\} - 1)$ grid blocks:

$$\mathbf{m}(i) = \sum_{j=1}^{l_{al}} \mathbf{y}((i-1) \times l_{al} + j) \times \hat{H}_l(j, 1) + \mathbf{e}(i) \quad \mod 2 \tag{10}$$

$$\mathbf{e}(i) = \sum_{k=i+1-\min\{l_{bl}, i\}}^{i-1} \sum_{j=1}^{l_{al}} \mathbf{y}((k-1) \times l_{al} + j) \times \hat{H}_l(j, i-k+1) \quad \mod 2 \tag{11}$$

where $\mathbf{e}(i)$ can be considered as the carry of the last block and is determined before generating the current block. As a result, the first column of $\hat{H}_l$ should be all ones so that Eq. (10) can be solved by modifying the only one dry element.

Then the selection of the height of $\hat{H}_l$, l_{bl} is analyzed. Message bits can be carried by the only one dry element in every l_{al} elements in the extreme case we considered. Thus the embedding encoder degenerates into an LSB replacement. Equation (10) has only one solution given $\mathbf{m}(1)$, leaving a fixed $\mathbf{e}(1)$ for the next embedding. Consequently, the second stego segment satisfying $\mathbf{e}(1) + \sum_{j=1}^{l_{al}} \mathbf{y}(j)$ mod $2 = \mathbf{m}(2)$ has only one solution given the index of the dry element. It is similar when embedding the rest message bits. As a result, the expected embedding cost is independent with the height of $\hat{H}_l$ and of value

$$D(\mathbf{x}', \mathbf{y}') = \frac{1}{2} \sum_{j \in \mathcal{J}_{\mathrm{dry}}} \mathbf{w}(j) \tag{12}$$

where $\mathcal{J}_{\mathrm{dry}}$ is the index set of all the dry elements. Nevertheless, the number of possible paths increases exponentially with the number of dry elements in a cover segment. In this case, using a larger l_{bl} can better ensurer the embedding near the optimal bound boundary [3,7]. As a result, we select l_{bl} according to the cover distribution. For rather large wet region, a small value, e.g., $l_{bl} = 2$ will suffice.

At last, we discuss the number of sub-submatrices in each $\hat{H}$. It is equals to l_a/l_{ah} for $\hat{H}_h$, which can be set with a large value to increase the embedding efficiency. Regarding to $\hat{H}_l$, a large l_a/l_{al} may waste the area suitable for the embedding. However, a too small l_a/l_{al} may raise embedding failures. From Eq. (10) it can be derived that

$$\mathbf{o}(i) = \sum_{j=(i-1) \times (l_a/l_{al})+1}^{i \times (l_a/l_{al})} \mathbf{m}(j) + \mathbf{e}(j) \tag{13}$$

Since $\mathbf{e}(1) = 0$, the probability of successfully embedding $\mathbf{m}(1)$ is only $1/2$ if $l_a/l_{al} = 1$. To deal with it, It should satisfy that $l_a/l_{al} \geq 2$, or the first cover segment processed by $\hat{H}_l$ could not be used to carry message bits. On the other hand, there should be at least one $\hat{H}_l$ processing cover elements that contain more than one dry element in a $\hat{H}$, otherwise Eq. (10) is high likely to have no solution given designated parities $\mathbf{o}(i)$ and message bits $\mathbf{m}((i-1) \times (l_a/l_{al}) + 1), \mathbf{m}((i-1) \times (l_a/l_{al}) + 2). \cdots$.

3.2   Mapping Between Parity and Embedding Rate

In Eq. (1), function $\Phi()$ maps the parity of a stego segment to the embedding rate used in that segment. It is used to allay the detectable influence incurred by designating parities. Consider a down-sampled image $X^{\downarrow}$, each of whose pixels is the sum of the pixels in the image segment at the same place in the originally sized image X.

$$X^{\downarrow}(i) = \sum_{j=(i-1)\times l_a+1}^{i\times l_a} X(j) \tag{14}$$

Then designating the parities of the segments in the original image is equivalent to an LSB replacement in this down-sampled version. In view of this, $\Phi()$ should satisfy: 1) $\Phi()$ can be generated at both the sender and the receiver without any side information of the cover image and the secret message. 2) $\Phi()$ should be random, otherwise it would result in regular LSBs in the down-sampled image indicating local wetness of the image. 3) $\Phi()$ should minimize the distortion caused by modifying LSBs of the down-sampled image, because a warder may train a steganalyzer on these down-sampled images to detect the proposed VR-STC.

Herein we meet the above criteria by means of STC. Briefly, we obtain $\Phi()$ by embedding encrypted indicators of embedding rates, ρ, in the down-sampled image by the STC. The generation of $\Phi()$ is as follows.

1. Generate down-sampled image $X^{\downarrow}$ from cover image X by Eq. (14).
2. Use certain distortion measurement to define the embedding cost of each pixel in $X^{\downarrow}$, yielding a cost map $W^{\downarrow}$.
3. Divide $X^{\downarrow}$ into vectors $\mathbf{x}^{\downarrow}$ of length l_{s1}, which is equivalent to combine each l_{s1} successive image segments in X into a group. Suppose there are l' vectors in total.
4. Generate a pseudorandom binary vector $\mathbf{r} \in \{0,1\}^{l'}$ by pre-negotiated encryption method and key.
5. For the i-th $\mathbf{x}^{\downarrow}$, segments in the cover image corresponding to the pixels in this vector should embed message bits with the same embedding rate. This embedding rate is determined based on the analysis in Sect. 3.1. Especially, ρ_l will be selected if the number of dry elements is no more than l_a/l_{ah} in some segments.
6. If ρ_l is selected, an indicator is saved as $\rho(i) = 1 - \mathbf{r}(i)$. This indicator will be used to give the parities each stego segments in the group should have, namely $\mathbf{o}((i-1)\times l_{s1}+1), \mathbf{o}((i-1)\times l_{s1}+2), ..., \mathbf{o}(i\times l_{s1})$, in the next step. Similarly, $\rho(i) = \mathbf{r}(i)$ will be saved if ρ_h is selected.
7. After obtaining a certain number of indicators, saying, l_{s2} indicators, they are considered as a message vector and embedded into $X^{\downarrow}$ by the STC with embedding rate $1/l_{s1}$ according to cost map $W^{\downarrow}$, yielding the parities of each stego segments $\mathbf{o}$:

$$\mathbf{o} = \text{STC}(\rho, X^{\downarrow} \mod 2, W^{\downarrow})$$

The above procedure gives us a $\Phi()$, namely the parity and the embedding rate associated with each stego segment. Then they are used to embed message bits by the parity-aware encoder. In practice, we set $1 < l_{s1} \leq 2$ to fully utilize the dry elements in image segments. Note that the length of each $\mathbf{x}^{\downarrow}$ does not need to be same. For example, we can set $l_{s1} = 1$ for the 1-st, 3-rd, 5-th $\mathbf{x}^{\downarrow}$, and $l_{s1} = 2$ for the 2-nd, 4-rd, and 6-th vectors. This is equivalent to embedding indicators with embedding rate 0.75. On the other hand, it calls for that all the segment in a group should be embedded with the same embedding rate. As a result, we have to reset the embedding rate of the first segment to ρ_l if we find that the second segment have to be embedded with ρ_l.

4   Experimental Results

4.1   Embedding Efficiency Loss

The VR-STC performs similar to the combination of two STC codes if we removed the parity limitation in Eq. (1). Sometimes the designated parity is inconsistent with that of the stego vector found by the original STC, which may lower the embedding efficiency. In this section we evaluate the embedding efficiency loss caused by the parity designation.

For the sake of simplification, the parity-aware encoder is only equipped with one sub-submatrix (i.e., $\rho_l = \rho_h = \rho$). We compare 4 parity designation strategies: 1) message bits are embedded using the original STC (denoted as Arb-parity), 2) the parities of all the stego segments are odd (denoted as Odd-parity), 3) the parities of all the stego segments are even (denoted as Even-parity), 4) the parities are set by using the procedure presented in Sect. 3.2 (denoted as Opt-parity). In each strategy, pseudorandom binary sequences of 480-bit length are employed as cover vectors, whose embedding costs are supposed to be constant $(\mathbf{w}(i) = 1)$, linear $(\mathbf{w}(i) = i)$, and square $(\mathbf{w}(i) = i^2)$. They are used to embed pseudorandom message bits at embedding rate $\rho \in \{1/6, 1/4, 1/3, 1/2\}$. The width of submatrix is set as $l_a = 12$ in Odd-parity, Even-parity and Opt-parity. Further, in Opt-parity, we set $l_{s1} = 2$ and $l_{s2} = 2$, and the embedding cost of each element in $\mathbf{x}^{\downarrow}$ is the average of the embedding costs of the dry elements in the corresponded segment in the originally sized cover vector.

Figure 3 compares the embedding distortion among different parity designation strategies. The result is averaged over 100 times simulations. It can be observed a constant gap between the VR-STC and the original STC, which seems independent with the embedding rate. Moreover, the Opt-parity strategy would further lower the embedding efficiency slightly. It may be because that more cover elements have to be flipped to preserve the LSBs of the down-sampled image. Nevertheless, the Opt-parity strategy performs best on the down-sampled vectors due to its targeted design.

The influence of submatrix' width, l_a, on the embedding efficiency loss is further evaluated. We fix the embedding rate as $\rho = 1/2$ and vary l_a from 20 to 60. Experimental results shown in Fig. 4 indicate that enlarge the width can allay the embedding efficiency loss, and reduce the embedding cost on the down-sampled vectors. However, a large l_a would weaken the adaptiveness of the proposed code on dynamic local cover distributions. Empirically, its value is set so that $l_a/l_{ah} \leq 4$.

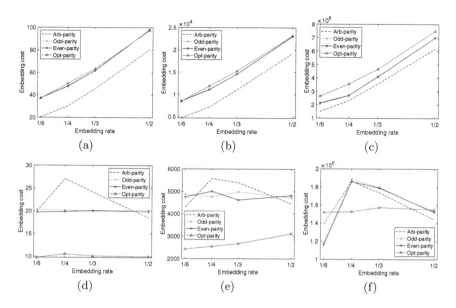

Fig. 3. Illustration of the embedding efficiency loss incurred by the VR-STC. The cover vector is associated with a) constant, b) linear, and c) square embedding cost. d), e), and f) show the embedding distortion on the down-sampled vectors, whose embedding costs are obtained by using the embedding costs in a), b), and c) respectively.

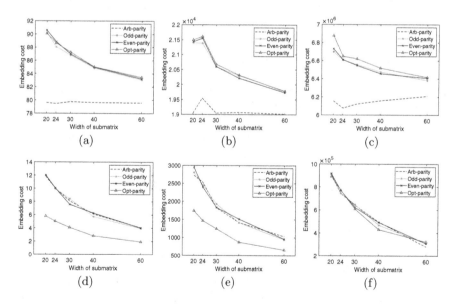

Fig. 4. Illustration of the influence of the width of submatrices on the embedding efficiency loss. The embedding costs are constant in a) and d), linear in b) and e), and square in c) and f). a)–c) show the results on the cover vectors, while d)–f) show the results on the down-sampled versions.

4.2   Performance on Artificial Bursty Channel

This section tests the adaptiveness of the VR-STC and original STC on bursty channels. The definition of the bursty channel is similar to the classic one in [11]. When referring to a $(prob, dura)$-channel, it means that the probability of an incoming element being wet is $prob$, and each time the channel will stay in wet for $dura$ elements. The VR-STC is constructed by two sub-submatrices providing embedding rates $\rho_h = 1/2$ and $\rho_l = 1/6$, and uses the Opt-parity strategy. Both VR-STC and original STC employ linear embedding cost. The other experimental settings are as same as those in Sect. 4.1.

Since the embedding rate keeps changing during the VR-STC's embedding procedure, we adjust the embedding rate of the original STC to be similar to the average embedding rate of the VR-STC given a cover vector. The considered settings of bursty channels, together with the averaged embedding rates provided by the VR-STC in each setting, are listed in Tabel 1.

Table 1. Averaged embedding rates of VR-STC under different channel settings.

$dura$	$prob$						
	0.1	0.2	0.3	0.4	0.5	0.6	0.7
3	240.00	192.64	181.20	132.72	99.92	94.88	90.64
5	199.28	142.56	116.16	91.44	82.00	82.80	82.08
7	149.28	116.16	105.68	81.16	80.04	80.08	80.04

The numbers of failed embedding attempts under different $(prob, dura)$-channels are compared in Figs. 5(a) to 5(c). It can be observed that the VR-STC can acquire stable and high successful rate when $dura \leq 6$. In order to achieve a similar embedding rate, the original STC has to use a fixed high embedding rate, resulting in high failure rate. On the other hand, the constructed VR-STC does not target at $dura > 6$, and thus performs poorly when $dura = 7$. We further compare the averaged embedding costs of the successful embedding. The comparison results shown in Figs. 5(d) to 5(f) indicate that the VR-STC increases the embedding cost slightly, which is coincides with the results in Sect. 4.1. Note that some points are absent in these figures due to the lack of successful embedding.

4.3   Performance on Binary Images

Binary images are frequently used in document images, handwritings, CAD graphs and so on, and many steganographic schemes on them have been developed [17,18]. Further, their nonstationary property is easy to exhibit. As a result, we compare the VR-STC and original STC on the binary images in the end of this paper. The dataset presented in [17], which consists of 5000 binary images of size 256×256, is employed. The VR-STC is constructed by two sub-submatrices

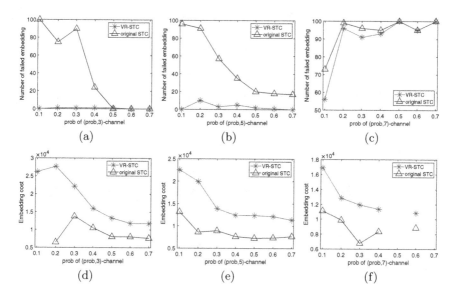

Fig. 5. a), b), c) show the comparison of numbers of failed embedding attempts. d), e), f) show the comparison of averaged embedding costs of successful embedding. In each figure we fix the *dura* of (*prob, dura*)-channel and vary *prob* from 0.1 to 1.

providing embedding rates $\rho_h = 1/2$ and $\rho_l = 1/8$. Both VR-STC and STC accept cover vectors of length 480, and the embedding rate of the original STC is adjusted to be similar to that of the VR-STC. In the embedding procedure, a cover image is divided into 2×4 sized blocks, and those containing different pixel values are used to fill cover vectors. The threshold and reembedding mechanisms suggested in [17] are employed. The embedding costs are computed by the measurement defined in [17] as well. Further, pixels with embedding costs larger than 8 are set with wet.

Herein we only compare the imperceptibility of different schemes, since stego binary images with better visual quality can usually provide stronger statistical undetectability [17]. The perceptual distortion on stego images is compared in Fig. 6. It suggests that the VR-STC can better avoid incurring distortion in flat region. The Opt-parity strategy further reduces the embedding distortion by the VR-STC on down-sampled images. Since the visual difference is subtle. We compare the imperceptibility by some objective measurements including Hamming distance, DRD [19], and ELD [20]. The comparison results listed in Table 2 support that the proposed scheme can generally preserve the visual quality better. This is because the bursty characteristic of binary images makes the VR-STC more suitable compared with the original STC.

Table 2. Comparison of averaged embedding distortion between VR-STC and STC.

	Averaged capacity	Original sized image			Down-sampled image		
		Hamming	DRD	ELD	Hamming	DRD	ELD
VR-STC	4.51%	**1120.42**	**1.8183**	**4249.55**	**405.88**	3.0153	**1631.01**
Original STC	4.51%	1161.95	1.8203	4529.64	411.62	**2.7129**	1631.68

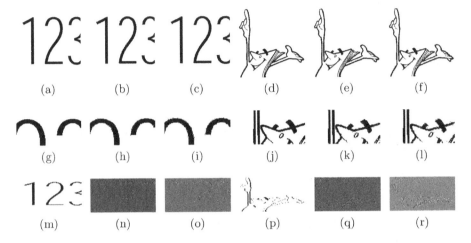

Fig. 6. Imperceptibility comparison on binary images. (a) and (d) are cover images. (b) and (e) are the corresponded stego images obtained by the VR-STC. (c) and (f) are stego images obtained by the original STC. (g)–(l) are the image segments cropped from (a)–(f), respectively. (m) and (p) are the down-sampled versions of (a) and (d), respectively. (n), (o), (q), and (r) are the difference maps between the down-sampled versions of the cover and corresponded stego images.

5   Conclusion

In this paper, we consider a special channel for covert communication, the bursty channel, and propose the variable rate syndrome-trellis codes (VR-STC) to fully utilize this channel. The VR-STC provides two alternative embedding rates to support this channel. Between them, the lower one is set to be low enough to pass through the wet region, and the higher one is used to provide embedding capacity. The stego vector is composed of a set of stego segments. Which embedding rate is used for a segment is indicated by the parity of that segment. The core design is the parity-aware encoder, which uses two sub-submatrices to generate two candidate stego segments with different embedding rates and opposite parities. Either of them can be used to form the final stego vector. We use the Viterbi algorithm to find the closed stego segments generated by the two sub-submatrices. Moreover, the selection of sub-submatrices is discussed. The mapping between parities and embedding rates is optimized to attenuate the embedding efficiency loss caused by the parity designation, meanwhile improving the statistical undetectability. Experimental results show that the proposed scheme is suitable for bursty channels. It can provide considerable stego image quality without the aid of random permutation. We also notice the relation between the VR-STC and the batch steganography [21,22]. The latter focuses on the capacity spreading strategy and pays less attention on the practical implementation details, such as the transmission of cover merging and selection parameters. Thus the VR-STC may help practically implement batch steganographic schemes. Nevertheless, the cover images considered there are more stationary and easier to deal with.

Acknowledgements. This work was supported by Key R&D Program of Guangdong Province (Grant No. 2019B010136003), National Natural Science Foundation of China (Grant No. 61802145), Natural Science Foundation of Guangdong Province, China (Grant No. 2019B010137005, 2017A030313390, 2018A030313387), Science and Technology Program of Guangzhou, China (Grant No. 201804010428), the Fundamental Research Funds for the Central Universities, the Opening Project of State Key Laboratory of Information Security, the Opening Project of Guangdong Key Laboratory of Intelligent Information Processing and Shenzhen Key Laboratory of Media Security.

References

1. Cachin, C.: An information-theoretic model for steganography. In: Aucsmith, D. (ed.) IH 1998. LNCS, vol. 1525, pp. 306–318. Springer, Heidelberg (1998). https://doi.org/10.1007/3-540-49380-8_21

2. Fridrich, J., Goljan, M., Lisonek, P., Soukal, D.: Writing on wet paper. IEEE Trans. Sig. Process. **53**(10), 3923–3935 (2005)

3. Filler, T., Judas, J., Fridrich, J.: Minimizing additive distortion in steganography using syndrome-trellis codes. IEEE Trans. Inf. Forensics Secur. **6**(3), 920–935 (2011)

4. Li, W., Zhang, W., Li, L., Zhou, H., Yu, N.: Designing near-optimal steganographic codes in practice based on polar codes. IEEE Trans. Commun. **68**(7), 3948–3962 (2020)

5. Sedighi, V., Cogranne, R., Fridrich, J.: Content-adaptive steganography by minimizing statistical detectability. IEEE Trans. Inf. Forensics Secur. **11**(2), 221–234 (2015)

6. Denemark, T., Fridrich, J.: Model based steganography with precover. Electron. Imaging **2017**(7), 56–66 (2017)

7. Köhler, O.M., Pasquini, C., Böhme, R.: On the statistical properties of syndrome trellis coding. In: Kraetzer, C., Shi, Y.-Q., Dittmann, J., Kim, H.J. (eds.) IWDW 2017. LNCS, vol. 10431, pp. 331–346. Springer, Cham (2017). https://doi.org/10.1007/978-3-319-64185-0_25

8. Dirfaux, F.: Key frame selection to represent a video. In: Proceedings 2000 International Conference on Image Processing (Cat. No. 00CH37101), Volume 2, pp. 275–278. IEEE (2000)

9. Kobayashi, M., Okabe, T., Sato, Y.: Detecting forgery from static-scene video based on inconsistency in noise level functions. IEEE Trans. Inf. Forensics Secur. **5**(4), 883–892 (2010)

10. Oh, S.M., Cho, S., Kim, J.H., Kwun, J.: VoIP scheduling algorithm for AMR speech codec in IEEE 802.16 e/m system. IEEE Commun. Lett. **12**(5), 374–376 (2008)

11. Forney, G.: Burst-correcting codes for the classic bursty channel. IEEE Trans. Commun. Technol. **19**(5), 772–781 (1971)
12. Park, J., Park, J., Bhunia, S.: VL-ECC: variable data-length error correction code for embedded memory in DSP applications. IEEE Trans. Circuits Syst. II: Express Briefs **61**(2), 120–124 (2013)
13. Namba, K., Lombardi, F.: Parallel decodable two-level unequal burst error correcting codes. IEEE Trans. Comput. **64**(10), 2902–2911 (2014)
14. Demirci, M., Reviriego, P., Maestro, J.A.: Unequal error protection codes derived from double error correction orthogonal Latin square codes. IEEE Trans. Comput. **65**(9), 2932–2938 (2015)
15. Shin, D., Park, J., Park, J., Paul, S., Bhunia, S.: Adaptive ECC for tailored protection of nanoscale memory. IEEE Des. Test **34**(6), 84–93 (2016)
16. Ker, A.D.: On the relationship between embedding costs and steganographic capacity. In: Proceedings of the 6th ACM Workshop on Information Hiding and Multimedia Security, pp. 115–120 (2018)
17. Feng, B., Lu, W., Sun, W.: Secure binary image steganography based on minimizing the distortion on the texture. IEEE Trans. Inf. Forensics Secur. **10**(2), 243–255 (2015)
18. Yeung, Y., Lu, W., Xue, Y., Huang, J., Shi, Y.Q.: Secure binary image steganography with distortion measurement based on prediction. IEEE Trans. Circuits Syst. Video Technol. **30**(5), 1423–1434 (2020)
19. Lu, H., Kot, A.C., Shi, Y.Q.: Distance-reciprocal distortion measure for binary document images. IEEE Sig. Process. Lett. **11**(2), 228–231 (2004)
20. Cheng, J., Kot, A.C.: Objective distortion measure for binary text image based on edge line segment similarity. IEEE Trans. Image Process. **16**(6), 1691–1695 (2007)
21. Ker, A.D., Pevny, T.: Batch steganography in the real world. In: Proceedings of the on Multimedia and Security, pp. 1–10 (2012)
22. Wang, Z., Zhang, X., Qian, Z.: Practical cover selection for steganography. IEEE Sig. Process. Lett. **27**, 71–75 (2020)

Steganographic Distortion Function for Enhanced Images

Zichi Wang, Guorui Feng, and Xinpeng Zhang[(⊠)]

Shanghai Institute for Advanced Communication and Data Science,
School of Communication and Information Engineering,
Shanghai University, Shanghai 200444, China
{wangzichi, xzhang}@shu.edu.cn, fgr2082@aliyun.com

Abstract. Contrast enhancement is widely used to improve the visual quality of images. This paper proposes a distortion function for steganography in enhanced images. The pixel prediction error and the cost of the pixel value itself are joint to fit the unique properties of enhanced image. Given a cover image, each pixel is predicted by its neighbors. The prediction error, which reflects the texture complexity, are combined with the cost of the pixel value itself to form the final distortion function. With the proposed distortion function, secret data can be embedded into the enhanced cover image with minimal image distortion using syndrome trellis coding (STC). As a result, less detectable artifacts left in the stego images, so that high undetectability is achieved.

Keywords: Steganography · Enhanced images · Distortion function

1 Introduction

Steganography is a technique to embed secret data into digital media to achieve covert communication [1–3]. At the initial stage, most steganographic methods decrease the number of modifications to guarantee undetectability [4–6]. However, the undetectability of steganography is also determined by the locations of modifications. As present, the most popular steganographic framework is based on the minimization of additive distortion [7], which is achieved by syndrome trellis coding (STC) [8]. In this framework, a distortion function is defined to assign embedding costs for all cover elements to quantify the modification-effect. Many distortion functions have been designed for natural image in spatial domain [9–12] and DCT domain [13–16].

For other kinds of images, new distortion functions should be developed to fit the unique properties. Contrast enhancement is widely used to improve the visual quality of images, especially the content details. Several original natural images and their enhanced versions are shown in Fig. 1. The enhanced images are produced by histogram equalization which is a popular approach to achieve contrast enhancement [17]. From Fig. 1, we can see that the enhanced images are more legible than the original ones. In the enhanced versions, the details are easy to identify. This is important to medical, legal, and astronomical applications. In addition, the correlation between pixels in enhanced images is different from that in original natural image, since the

© Springer Nature Switzerland AG 2021
X. Zhao et al. (Eds.): IWDW 2020, LNCS 12617, pp. 31–40, 2021.
https://doi.org/10.1007/978-3-030-69449-4_3

histograms of enhanced images have been extruded. So, the properties of enhanced images are quite different with original natural images.

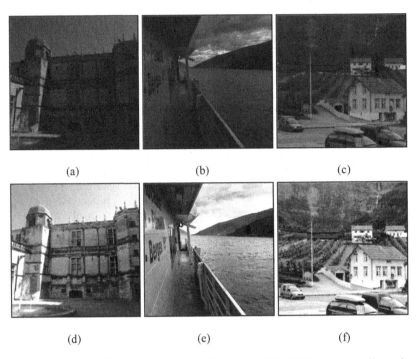

(a)   (b)   (c)

(d)   (e)   (f)

Fig. 1. Demonstration of (a)–(c) several natural images and (d)–(f) the corresponding enhanced versions.

Existing distortion functions [9–16] aim to restrain the modifications caused by embedding into texture and complex regions to conceal the modification trace. Although these distortion functions perform well in original natural image, they are still unsuitable for enhanced image since the unique properties of enhanced image have not been considered. It is shown in [18] that the undetectability of existing steganographic methods will decreased clearly when the cover image has been enhanced. The modification-trace caused by existing steganographic methods can be discovered by modern steganalytic tools easily when the cover images are enhanced. Therefore, it is necessary to develop customized distortion function for enhanced image. To the best of our knowledge, the distortion function designed for enhanced image has not been reported in the literature.

To this end, we firstly propose a distortion function for steganography in enhanced image. The pixel prediction error and the cost of the pixel value itself are joint to fit the unique properties of enhanced image. Given a cover image, each pixel is predicted by its neighbors. The prediction error, which reflects the texture complexity, are combined with the cost of the pixel value itself to form the final distortion function. When secret data is embedded with the proposed distortion function, the obtained stego image lefts less detectable artifacts.

2 Proposed Distortion Function

The structure of the proposed method is shown in Fig. 2. Given a cover image, each pixel is predicted to form a predicted image. Meanwhile, the grayscale cost of each pixel is employed to measure the embedding cost of gray level itself. Finally, the predicted image and grayscale costs are used to calculate the final embedding cost of each pixel.

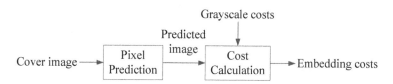

Fig. 2. Sketch of the proposed method.

2.1 Pixel Prediction

For an image $\mathbf{X}$ sized $M \times N$, denote the (i, j)-th pixel as $x(i, j)$, where $i \in \{1, 2, ..., M\}$, $j \in \{1, 2, ..., N\}$. To predict $x(i, j)$, the pixels $x(i + 1, j)$, $x(i - 1, j)$, $x(i, j + 1)$, and $x(i, j - 1)$ are employed. The nonexistent pixel which is out of the image boundary would be obtained by pixel symmetric padding. For example, $x(i + 1, j)$ can be obtained by copying $x(i - 1, j)$ when it is out of the image boundary, and vice versa.

Firstly, the vertical residual $d_v(i, j)$ and horizontal residual $d_h(i, j)$ of $x(i, j)$ are calculated using Eqs. (1) and (2). The reason for the using of quadratic difference is that the correlation between pixels in natural image is nonlinear. It is the same with enhanced natural image.

$$d_v(i,j) = [x(i-1,j) - x(i+1,j)]^2 \tag{1}$$

$$d_h(i,j) = [x(i,j-1) - x(i,j+1)]^2 \tag{2}$$

Then, the rates $r_v(i, j)$ and $r_h(i, j)$ of the two kinds of residuals are calculated using Eqs. (3) and (4) to measure the weights in vertical and horizontal directions. Where $\varepsilon = 10^{-8}$ is used to avoid the value of denominator becoming zero, and guarantee that the summation of $r_v(i, j)$ and $r_h(i, j)$ is equal to 1. In addition, it can be seen that the vertical rates $r_v(i, j)$ is determined by horizontal residual $d_h(i, j)$, and the horizontal rates $r_h(i, j)$ is determined by vertical residual $d_v(i, j)$. The rationale is that a larger $d_h(i, j)$ means weaker correlation in horizontal direction, so that the vertical weight should be larger. In other words, $r_v(i, j)$ is in inverse proportion to $d_h(i, j)$. Conversely, $r_h(i, j)$ is in inverse proportion to $d_v(i, j)$.

$$r_v(i,j) = \frac{d_h(i,j) + \varepsilon}{d_v(i,j) + d_h(i,j) + 2\varepsilon} \tag{3}$$

$$r_h(i,j) = \frac{d_v(i,j) + \varepsilon}{d_v(i,j) + d_h(i,j) + 2\varepsilon} \tag{4}$$

Finally, pixel $x(i, j)$ is predicted into $\hat{x}(i,j)$ using $x(i + 1, j)$, $x(i - 1, j)$, $x(i, j + 1)$, and $x(i, j - 1)$ as shown in Eq. (5). Where $w_v(i, j)$ and $w_h(i, j)$ are the weights in vertical and horizontal directions, which are defined in Eqs. (6) and (7). Since vertical and horizontal texture complexity of different images are different, a parameter $\alpha \in [0, 1]$ is used to adjust the ratio of the weights in vertical and horizontal directions. Furthermore, the number 4 of denominator in Eqs. (6) and (7) are used to normalize the summation of the four weights to 1.

$$\hat{x}(i,j) = w_v(i,j) \cdot x(i+1,j) + w_v(i,j) \cdot x(i-1,j) + w_h(i,j) \cdot x(i,j+1) + w_h(i,j) \cdot x(i,j-1) \tag{5}$$

$$w_v(i,j) = \frac{r_v(i,j) + \alpha}{4} \tag{6}$$

$$w_h(i,j) = \frac{r_h(i,j) + 1 - \alpha}{4} \tag{7}$$

To determine the value of α, eleven candidate values $\{0, 0.1, 0.2, \ldots, 1\}$ are employed for pixel prediction respectively to obtain eleven predicted images. Each predicted image is compared with the given image $\mathbf{X}$, and the minimal MSE (Mean Square Error) e is calculated using Eq. (8). Then, the value of α corresponding to the minimal e is used as the final α.

$$e = \frac{1}{MN} \sum_{i=1}^{M} \sum_{j=1}^{N} [\hat{x}(i,j) - x(i,j)]^2 \tag{8}$$

2.2 Cost Calculation

With the predicted pixel $\hat{x}(i,j)$, the cost $\rho(i, j)$ of modifying $x(i, j)$ can be calculated according to the difference between $x(i, j)$ and $\hat{x}(i,j)$. Since larger difference value means weaker spatial correlation of pixels and means lower embedding cost, the cost value should be in inverse proportion to difference value. So, $\rho(i, j)$ can be calculated using Eq. (9).

$$\rho(i,j) = \frac{\rho_g(x(i,j))}{|x(i,j) - \hat{x}(i,j)| + \varepsilon} \tag{9}$$

where $\rho_g(z)$, $z \in \{0, 1, ..., 255\}$ is the cost of the grayscale value of pixel $x(i, j)$ itself, which is proposed in [19], as shown in Eq. (10).

$$\rho_g(z) = \frac{1}{\left| z - 255(z/255)^{1.001} \right|} \tag{10}$$

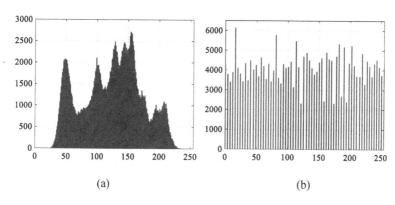

(a)   (b)

Fig. 3. Histogram comparison between (a) a natural image and (b) its enhanced version using histogram equalization.

We employ this grayscale cost to enrich our distortion function. The reason is that the spatial correlation of the enhanced image is different with natural image. The distribution of grayscale values of enhanced image is wider than that of natural image, as shown in Fig. 3. At this case, the histogram of correlations of enhanced image is weaker than that of natural image. Therefore, the embedding cost of gray level itself is necessary for enhanced image. The process of distortion function design can be summarized as the following algorithm.

ALG.1 Distortion function calculation

Input: Enhanced image **X**

Output: Embedding costs $\rho(i, j)$

---

1) Calculate the predicted pixel value $\hat{x}(i, j)$ for each cover pixel $x(i, j)$ using Equation (5);

2) Calculate the cost of the grayscale value for each cover pixel $x(i, j)$ using Equation (10);

3) Obtain the final cost value for each cover pixel $x(i, j)$ using Equation (9).

After the embedding cost values are obtained, secret data can be embedded into cover image X using the STC framework which minimizes the additive distortion between X and its stego version.

3 Experimental Results

In this section, some experiments are conducted to verify the effectiveness of the proposed distortion function. Firstly, we setup the experimental environments. We then provide a study of undetectability of our method when compared to the current state of the art.

3.1 Experiment Setup

The 1338 images sized 512×384 in UCID [20] and the 10000 images sized 512×512 in BOSSbase ver. 1.01 [21] are processed with histogram equalization (achieved by the function histeq(·) of MATLAB) to produce the enhanced images which are then used as cover images for embedding.

To verify the effectiveness of the proposed method, we compare our method with the popular distortion functions HILL [11], SUNIWARD [10] and WOW [9]. All embedding tasks are done by the embedding simulator [22] since it is widely used to simulate the optimal embedding. The payloads are set as 0.001, 0.002, 0.003, 0.004, and 0.005 bpp (bits per pixel), respectively.

For steganlaysis, the popular feature extraction methods SRMQ1 [23] and SPAM [24] are used to extract the feature sets of cover and stego images. Then, the ensemble classifier [25] is used to measure the property of feature sets, which is widely used for steganalysis. One-half of the cover and stego feature sets are used for training, while the remaining sets are used for testing. Then the criterion to evaluate the undetectability performance of steganography can be P_E which is the minimal total error with identical priors achieved on the testing sets, as shown in Eq. (11).

$$P_E = \min_{P_{FA}} \left(\frac{P_{FA} + P_{MD}}{2} \right) \tag{11}$$

where and P_{FA} and P_{MD} are the false alarm rate and missed detection rate respectively. A high value of P_E means high undetectability.

3.2 Image Quality

The demonstrations of images before and after embedding are shown in Fig. 4. Where Fig. 4(a) is a natural image without enhanced. Its enhanced version is shown in Fig. 4 (b) which is used as cover. After embedded by our method with payload 0.001, 0.002, 0.003, and 0.004 bpp respectively, the obtained stego images are shown in Fig. 4(c), (d), (e) and (f) correspondingly.

It can be seen that the stego images are close to the cover version, which means the visual quality of the stego images is satisfactory regardless of the payload. Therefore,

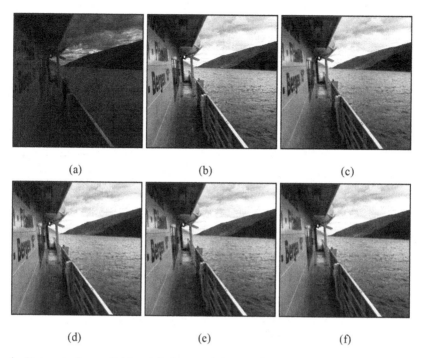

Fig. 4. Demonstrations of (a) original natural image, (b) enhanced cover image and the corresponding stego images using our method with payload (c) 0.001 bpp, (d) 0.002 bpp, (e) 0.003 bpp, and (f) 0.004 bpp.

the nice quality of enhanced image is reserved after embedded using our method. In other words, our method decreases the usability of enhanced image very slightly.

3.3   Undetectability Comparison

The undetectability comparisons between our method and other distortion functions against SPAM and SRMQ1 tested on UCID and BOSSbase ver. 1.01 are shown in Fig. 5 and Fig. 6 respectively.

It is clear that the undetectability performance of our method is better than other distortion functions for all cases, regardless of the steganalytic tools and payload. For image set UCID, the improvement on P_E by our method compared with HILL for payload 0.001 bpp against SRMQ1 is 3.54%. Compared with SUNIWARD and WOW, the improvements are 3.63% and 3.25% respectively for the same cases. For the cases against SPAM, the improvement on P_E by our method compared with HILL, SUNIWARD and WOW for payload 0.001 bpp are 3.30%, 3.45%, and 2.95% respectively. For image set BOSSbase ver. 1.01, similar results can be observed. Compared with HILL, the improvement on P_E by our method for payload 0.001 bpp against SRMQ1 is 5.40%. Compared with SUNIWARD and WOW, the improvements are 6.36% and 5.45% respectively for the same cases. The improvement on undetectability is because

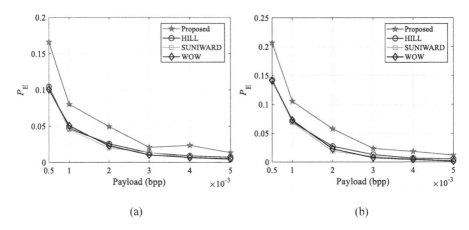

Fig. 5. Undetectability comparisons between the proposed method and several existing distortion functions on UCID against (a) SRMQ1 and (b) SPAM.

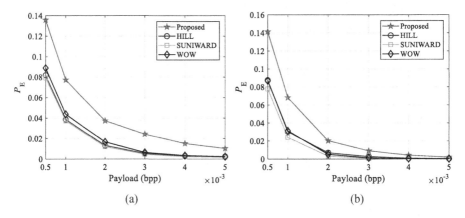

Fig. 6. Undetectability comparisons between the proposed method and several existing distortion functions on BOSSbase ver. 1.01 against (a) SRMQ1 and (b) SPAM.

that our method is designed to fit the unique properties of enhanced image, while the others not.

4 Conclusions

A distortion function for steganography in enhanced image is proposed in this paper. The pixel prediction error and the cost of the pixel value itself are joint to fit the unique properties of enhanced image. With the proposed distortion function, secret data can be embedded into the enhanced cover image with less detectable artifacts left in the stego images compared with existing distortion functions. Experimental results show that the undetectability performance of the proposed scheme is better than existing steganographic methods when check by modern steganalytic tools. We believe that the proposed scheme would be of some help to the development of digital image steganography, and more steganographic methods for special kinds of image can be developed in the future.

Acknowledgement. This work was supported in part by Natural Science Foundation of China (Grant 62002214), and Natural Science Foundation of Shanghai (Grant 19ZR1419000).

References

1. Li, S., Zhang, X.: Towards construction based data hiding: from secrets to fingerprint images. IEEE Trans. Image Processing **28**(3), 1482–1497 (2019)
2. Wang, Z., Zhang, X., Yin, Z.: Joint cover-selection and payload-allocation by steganographic distortion optimization. IEEE Signal Process. Lett. **25**(10), 1530–1534 (2018)
3. Tao, J., Li, S., Zhang, X., Wang, Z.: Towards robust image steganography. IEEE Trans. Circ. Syst. Video Technol. **29**(2), 594–600 (2019)
4. Fridrich, J., Soukal, D.: Matrix embedding for large payloads. IEEE Trans. Information Forensics and Security **1**(3), 390–395 (2006)
5. Zhang, X., Wang, S.: Efficient steganographic embedding by exploiting modification direction. IEEE Commun. Lett. **10**(11), 781–783 (2006)
6. Zhang, W., Zhang, X., Wang, S.: Maximizing steganographic embedding efficiency by combining hamming codes and wet paper codes. In: Solanki, K., Sullivan, K., Madhow, U. (eds.) IH 2008. LNCS, vol. 5284, pp. 60–71. Springer, Heidelberg (2008). https://doi.org/10.1007/978-3-540-88961-8_5
7. Fridrich, J., Filler, T.: Practical methods for minimizing embedding impact in steganography. In: Proceedings of SPIE, Security, Steganography, and Watermarking of Multimedia Contents IX, San Jose, CA, February 2007, pp. 2–3 (2007)
8. Filler, T., Judas, J., Fridrich, J.: Minimizing additive distortion in steganography using syndrome-trellis codes. IEEE Trans. Inf. Forensics Secur. **6**(3), 920–935 (2011)
9. Holub, V., Fridrich, J.: Designing steganographic distortion using directional filters. In: Proceedings of IEEE International Workshop on Information Forensics and Security, Binghamton, NY, USA, December 2012, pp. 234–239 (2012)
10. Holub, V., Fridrich, J., Denemark, T.: Universal distortion function for steganography in an arbitrary domain. EURASIP J. Inf. Secur. **2014**(1), 1–13 (2014)
11. Li, B., Wang, M., Huang, J., Li, X.: A new cost function for spatial image steganography. In: Proceedings of IEEE International Conference on Image Processing, Paris, France, October 2014, pp. 4206–4210 (2014)
12. Wang, Z., Lv, J., Wei, Q., Zhang, X.: Distortion function for spatial image steganography based on the polarity of embedding change. In: Shi, Y.Q., Kim, H.J., Perez-Gonzalez, F., Liu, F. (eds.) IWDW 2016. LNCS, vol. 10082, pp. 487–493. Springer, Cham (2017). https://doi.org/10.1007/978-3-319-53465-7_36
13. Guo, L.J., Ni, J.Q., Su, W.K., Tang, C.P., Shi, Y.Q.: Using statistical image model for JPEG steganography: uniform embedding revisited. IEEE Trans. Inf. Forensics Secur. **10**(12), 2669–2680 (2015)
14. Wang, Z., Zhang, X., Yin, Z.: Hybrid distortion function for jpeg steganography. J. Electron. Imaging **25**(5), 050501 (2016)
15. Wei, Q., Yin, Z., Wang, Z., Zhang, X.: Distortion function based on residual blocks for jpeg steganography. Multimed. Tools Appl. **77**(14), 17875–17888 (2018)

16. Wang, Z., Qian, Z., Zhang, X., Yang, M., Ye, D.: On improving distortion functions for JPEG steganography. IEEE Access **6**(1), 74917–74930 (2018)
17. Stark, J.A.: Adaptive image contrast enhancement using generalizations of histogram equalization. IEEE Trans. Image Process. **9**(5), 889–896 (2000)
18. Wang, Z., Zhang, X., Qian, Z.: Practical cover selection for steganography. IEEE Signal Process. Lett. **27**(1), 71–75 (2020)
19. Wei, Y., Zhang, W., Li, W., Yu, N., Sun, X.: Which gray level should be given the smallest cost for adaptive steganography. Multimed. Tools Appl. **77**, 17861–17874 (2018)
20. Schaefer, G., Stich, M.: UCID - an uncompressed colour image database. In: Proceedings of Conference on Storage and Retrieval Methods and Applications for Multimedia, San Jose, CA, USA, January 2004, pp. 472–480 (2004)
21. Bas, P., Filler, T., Pevný, T.: "Break our steganographic system": the ins and outs of organizing BOSS. In: Filler, T., Pevný, T., Craver, S., Ker, A. (eds.) IH 2011. LNCS, vol. 6958, pp. 59–70. Springer, Heidelberg (2011). https://doi.org/10.1007/978-3-642-24178-9_5
22. Pevný, T., Filler, T., Bas, P.: Using high-dimensional image models to perform highly undetectable steganography. In: Böhme, R., Fong, P.W.L., Safavi-Naini, R. (eds.) IH 2010. LNCS, vol. 6387, pp. 161–177. Springer, Heidelberg (2010). https://doi.org/10.1007/978-3-642-16435-4_13
23. Fridrich, J., Kodovsky, J.: Rich models for steganalysis of digital images. IEEE Trans. Inf. Forensics Secur. **7**(3), 868–882 (2012)
24. Pevny, T., Bas, P., Fridrich, J.: Steganalysis by subtractive pixel adjacency matrix. IEEE Trans. Inf. Forensics Secur. **5**(2), 215–224 (2010)
25. Kodovsky, J., Fridrich, J., Holub, V.: Ensemble classifiers for steganalysis of digital media. IEEE Trans. Inf. Forensics Secur. **7**(2), 432–444 (2012)

Improving Text-Image Matching with Adversarial Learning and Circle Loss for Multi-modal Steganography

Yuting Hu[1,2], Han Cao[1], Zhongliang Yang[1,2(✉)], and Yongfeng Huang[1,2]

[1] Department of Electronic Engineering, Tsinghua University, Beijing 100084, China
huyt16@mails.tsinghua.edu.cn, {caoh16,yangzl15}@tsinghua.org.cn,
yfhuang@mail.tsinghua.edu.cn
[2] Beijing National Research Center for Information Science and Technology,
Beijing 100084, China

Abstract. This paper proposes a multi-modal steganography method based on an improved text-image matching algorithm. At present, most of the steganography methods are based on single modality of carriers and embed confidential information into the carriers by cover modification or cover synthesis. Since the distortions between the covers after embedding and the original covers are inevitable, these steganography methods can be detected by the existing steganalysis methods. To solve this problem, we propose multi-modal steganography which hides the confidential information in the semantic relevancy between two modalities of original carriers. The semantic relevancy between the two modalities is measured by text-image matching, which affects the imperceptibility of the proposed method to a large extent. In order to increase the security of multi-modal steganography, we improve the current text-image matching algorithm with adversarial learning and circle loss. By selecting and transmitting the original multi-modal carriers with high relevancy, the proposed method can escape from the detection of current steganalysis methods. It is also illustrated by the theoretical analysis and experimental results that the semantic relevancy between the selected multi-modal carriers is enhanced.

Keywords: Multi-modal steganography · Text-image matching · Adversarial learning · Circle loss

1 Introduction

Steganography is a technique of communicating secret messages without being noticed by a third party. With the popularity of the multimedia in network transmission, steganography hides the secret messages into the digital carriers such as images and texts and so on.

This research is supported by the National Key R&D Program (2018YFB0804103) and the National Natural Science Foundation of China (No. U1705261 and No. U1836204).

Currently, most of the existing steganography methods are based on single modality of carriers. Many steganography methods conceal secret messages by modifying the content of the cover [2,19]. But the modification traces left on the modified covers will make steganalysis possible [5,14]. With the development of deep learning technologies such as recurrent neural networks and generative adversarial networks, some steganography methods based on cover synthesis are put forward [3,15,18]. However, these kind of steganography methods can be detected by the steganalysis methods since the differences of the distribution between the synthesized covers and the original covers are inevitable [16,17]. In order to escape from the detection of the steganalysis methods, a kind of new steganography methods named coverless information hiding is proposed [20–23]. In these methods, original covers which contain the secret messages are selected from a constructed database and transmitted to the receiver. Nevertheless, the covers selected by the most coverless information hiding methods are content-irrelevant, which is not consistence with the general phenomena on the network transmission. Therefore, this kind of steganography is likely to detected by side channel steganalysis [8].

In order to resist the existing steganalysis methods, we propose a multi-modal steganography method based on an improved text-image matching algorithm. It can be regarded as an extension version of the previous work of our research group [6]. In the previous work, we propose a basic multi-modal steganography (MM-Stega) framework which utilizes a visual semantic embedding (VSE) model [4] to measure the relevancy between the two modalities. But there is constantly room for development. In this paper, we improve the text-image matching algorithm from two aspects. One is to pull close the distributions of the two modalities by utilizing adversarial learning. Another is to make the optimization flexible by replacing the commonly-used triplet loss with circle loss. Then we employee a two-stage procedure to train the text-image matching model with a large number of text-image pairs in the Twitter100k dataset [7]. In this way, the relevancy between the two modalities can be enhanced, which brings about higher security of the multi-modal steganography method.

The advantages of the proposed multi-modal steganography are two fold. First, since the secret messages are embedded in the relevancy between the two modalities, the covers after embedding are still original. Therefore, steganalysis methods based on the distortion between the original covers and the covers after embedding are unable to detect the proposed steganography method. Second, because the relevancy between the two modalities are measured with an improved text-image matching algorithm with adversarial learning and circle loss, the selected covers for covert communication are relevant in semantics. Thus, the proposed method can also have good resistance to side channel steganalysis which are based on the correlation between the images sequence [8].

The rest of this paper is organized as follows. Section 2 introduces related work about the baseline text-image matching model and two main loss functions for text-image matching. The proposed multi-modal steganography method is described in detail in Sect. 3. The experimental results and analysis are presented in Sect. 4. Finally, Sect. 5 concludes the paper.

2   Related Work

2.1   The Baseline Text-Image Matching Model

Visual semantic embedding (VSE) model [4] is adopted in the previous work of multi-modal steganography [6], which is also the baseline text-image matching model in this paper.

VSE model is made up of a text-embedding network and an image-embedding network. The text and image embedding networks can project the two modalities into a common space, where the embeddings of the two modalities which are relevant in semantics are close in distance.

Text-Embedding Network. Recurrent neural network (RNN) has been employed wildly in many natural language processing tasks. Each RNN has a recurrent hidden state whose activation at each time is dependent on the input at current time and the activations of the previous time. By this way, all the information of a sequence is encoded. Gated recurrent unit (GRU) [1] is a variant of RNN, which adds an update gate and a reset gate to solve the gradient explosion/vanishing problem of RNN.

In the text-embedding network, each word in the sentence is first converted into a 300-dim word vectors with a trainable look-up table. Afterwards, a one-layer GRU is employed to encode the sequence of the word vectors. The output of the hidden state at the last time step is used as the output of the GRU, which is a 1024-dim vector. At last, the text embedding u is normalized using its l_2 norm.

Image-Embedding Network. VGG19 [11] has broad application in numerous tasks of computer vision. It is adopted as the backbone of the image embedding network. The model pretrained on the ImageNet [10] is used for parameter initialization except for the final fully connected (FC) layer.

In the image-embedding network, the input image is resized to $224 \times 224 \times 3$ and the output of the image embedding network is a 1,024-dim vector. Finally, the image embedding v is normalized with its l_2 norm.

2.2   Loss Functions

Triplet Loss. Triplet loss is a common ranking loss used in the works of text-image matching [4,7,9]. It is defined as follows:

$$l(i, c) = \max_{c'}[\alpha - s(i, c) + s(i, c')]_+ + \max_{i'}[\alpha - s(i, c) + s(i', c)]_+ \qquad (1)$$

in which $[x]_+ = max(x, 0)$, α is a margin, c' denotes the hardest negative text for the query image i and i' denotes the hardest negative image for the query text c.

Circle Loss. Circle loss is proposed recently for deep feature learning, aiming to maximize the within-class similarity s_p and minimize the between-class similarity s_n [12]. Different from the majority of loss functions such as the triplet loss which seek to reduce $(s_n - s_p)$, circle loss re-weights each similarity score to highlight the less-optimized similarity scores by optimizing $(\alpha_n s_n - \alpha_p s_p)$. Circle loss is named due to its circular decision boundary $\alpha_n s_n - \alpha_p s_p = m$. By this way, circle loss benefits the deep feature learning with flexible optimization and definite convergence target. Circle loss is formulated by:

$$\mathcal{L}_{circle} = log[1 + \sum_{j=1}^{L} exp(\gamma \alpha_n^j (s_n^j - \Delta_n)) \sum_{i=1}^{K} exp(\gamma \alpha_p^i (s_p^i - \Delta_p))], \qquad (2)$$

where γ is a scale factor, Δ_n and Δ_p are the between-class and within-class margins, respectively. K and L are the number of within-class similarity scores and between-class similarity scores, respectively.

Experiments have demonstrated that the superiority of the circle loss on many deep feature learning tasks such as image retrieval, face recognition and person re-identification.

3   The Proposed Multi-modal Steganography

In this section, we will illustrate the proposed multi-modal steganography based on an improved text-image algorithm. The text-image matching model utilized in this paper is an refined version of the baseline text-image matching model [4] using adversarial learning [9] and circle loss [12]. The text-image matching model is trained on the Twitter100k dataset [7], which is a public large-scale text-image dataset. With the text-image matching model, the two modalities can be projected into a common space, where the relevancy between them can be measured directly by the distance of their embeddings. The text database and image database required in our method are constructed by selecting a number of texts and images from the Twitter100k dataset [7], which doesn't need much effort. The flow chart of our method for covert communication is represented in Fig. 1.

To convey the secret message, the sender first converts the secret message into a bit stream and divides it into n segments with the equal length k, where n and k are set according to the amount of the secret message and k is shared by the sender and receiver. Then each binary segment is transformed to a decimal integer. Thus, the secret message is converted to n decimal integers. Afterwards, a text is selected from the text database randomly. The relevancies between the selected text and the images in the image database are computed with the well-trained text-image matching model. Then all the images are sorted on the basis of the relevancies. After that, n decimal integers serve as the sort indexes to select n images from the image database. Each image can be selected at most once. After an image is selected, the sort indexes of the images behind the selected images are reduced by one. Finally, the selected text and n images are transmitted to the receiver.

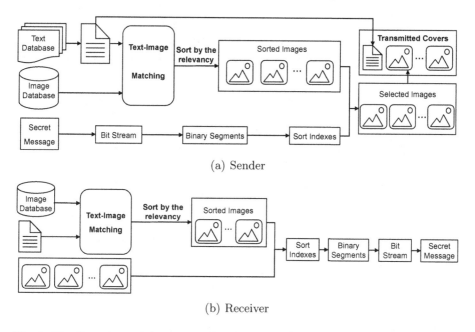

(a) Sender

(b) Receiver

Fig. 1. The flow chart of the proposed multi-modal steganography for covert communication at (a) sender and (b) receiver.

At the receiver end, the same text-image matching model and image database are shared with the sender. After receiving the text, the relevancies between the received text and the images in the image database are computed with the shared text-image matching model. Afterwards, all the images are sorted according to the relevancies. By matching the received images with the sorted images, the sort indexes of the received images can be obtained, which are n decimal integers. Then each decimal integer is converted into a binary sequence. After splicing the binary sequences of the n decimal integer together into a bit stream, the secret message can be recovered by converting the bit stream.

To conclude, the proposed multi-modal steganography has three main parts, *i.e.*, the text-image matching model, information hiding algorithm and information extraction algorithm.

3.1   The Text-Image Matching Model

We utilize the VSE model [4] as the baseline text-image matching model in our method. The VSE model [4] consists of a text embedding network and an image embedding network, as introduced in Sect. 2.1. Text and image can be transformed into the 1024-dim vectors u and v with the text embedding network and the image embedding network, respectively.

In the previous work, the VSE model is optimized using triplet loss whose formula is given in Sect. 2.2. However, as illustrated in [12], using triplet loss lacks flexibility for optimization and leads to ambiguous convergence status. In order to achieve a superior optimization for text-image matching, we replace the triplet loss with a novel circle loss which is introduced in Sect. 2.2. The similarity score in Eq. (2) is the inner product of the embeddings of text and image,

$$s(i, c) = v \cdot u. \tag{3}$$

Assume there are m pairs of texts and images in the training dataset. The text embedding matrix is made up of m 1024-dim text vectors and the image embedding matrix is formed of m 1024-dim image vectors. Then we can obtain a similarity matrix S by computing the product of the text embedding matrix and the image embedding matrix,

$$S = \begin{bmatrix} u_1^T \\ \vdots \\ u_m^T \end{bmatrix} \begin{bmatrix} v_1 \dots u_{n_i} \end{bmatrix} = \begin{bmatrix} s_{11} \dots s_{1m} \\ \vdots \ddots \vdots \\ s_{m1} \dots s_{mm} \end{bmatrix}. \tag{4}$$

In Eq. (4), the diagonal elements $\{s_{11}, ..., s_{mm}\}$ are within-class similarity scores while the rest elements $\{s_{ij}\}_{i \neq j}$ are between-class similarity scores. As suggested in [12], the circle loss is computed by substituting all the similarity scores in the similarity matrix into the Eq. (2). However, it is found by our experiment that the performance is unsatisfactory in this way. To address this problem, we refine the computation as follows. The circle loss is calculated with the similarity scores of each row of the similarity matrix. Then the overall circle loss is the sum of these circle losses, that is,

$$\mathcal{L}_{circle,i} = log[1 + \sum_{j \neq i} exp(\gamma \alpha_n^j (s_{ij} - \Delta_n)) exp(\gamma \alpha_p^i (s_{ii} - \Delta_p))], \tag{5}$$

$$\mathcal{L}_{circle} = \sum_{i=1}^{m} \mathcal{L}_{circle,i}. \tag{6}$$

Besides circle loss, we also utilize adversarial learning to further improve the text-image matching model. As illustrated in [9], the embedding distributions of the two modalities have sensible discrepancy when VSE model [4] is used. In order to pull close the embedding distributions of the two modalities in the common space, a modality classification network is added to the baseline VSE model for adversarial learning. The overall network structure of the text-image matching model is illustrated in Fig. 2.

The modality classification network contains a three-layer fully-connected network (FCN) to classify the embeddings of VSE into either image or text modality. We use the same output sizes of the three FC layers as [9], i.e., 1,024, 256, and 2, respectively. The first two layers employee ReLU as the non-linear activation and the last layer adopts the softmax function. The adversarial loss is the modality classification loss which is a cross-entropy loss.

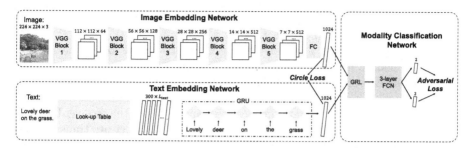

Fig. 2. The overall network structure of the improved text-image matching model, which consists of a text-embedding network, an image-embedding network and a modality classification network. Circle loss and adversarial loss are used to optimize the networks.

For the purpose of optimizing the circle loss and the adversarial loss simultaneously, a gradient reversal layer (GRL) is added before the FCN. Therefore, the total loss is defined as belows,

$$\mathcal{L} = \mathcal{L}_{circle} - \lambda \mathcal{L}_{adversarial} \tag{7}$$

where λ is a factor controlling the weight of the adversarial loss, which is set to 0.1 in the experiment.

Since there are three networks to be optimized in our method, we use a two-stage training procedure. In the first stage, only the parameters of the image-embedding and the text-embedding networks are updated by minimizing the circle loss as Eq. (6). In the second stage, the parameters of the modality classification network are updated when circle and adversarial loss are optimized simultaneously as Eq. (7).

3.2   Information Hiding Algorithm

There are five steps in the information hiding procedure, which will be introduced in detail as follows.

Step 1: Text selection. A text is selected randomly from the text database.
Step 2: Image sorting. The relevancies between the selected text and all the images in the image database are measured with the text-image matching model. All the images are sorted according to the relevancy in descending order.
Step 3: Secret message preprocessing. The secret message is converted into a bit stream and divided into n binary segments with the equal length k. n is the number of the companied images to the selected text. k is the number of bits which are represented by an image. n and k are set according to the amount of the secret message, the size of the image database and the limitations of the data transmission platform. Afterwards, each binary segment is transformed to a decimal integer. In this way, the secret message is represented by n

decimal integers, $[i_1, i_2, ..., i_n]$. The maximum and the minimum of the decimal integers are $2^k - 1$ and 0, respectively.

Step 4: Image selection. The n decimal integers serve as the sort indexes of the image to choose images from the sorted images. After an image is selected, the sort indexes of the images behind the selected images are reduced by one. Therefore, no repeated images will appear in the selected images.

Step 5: The selected text and n images are transmitted to the receiver by public transmission channels such as Twitter and Weibo or private transmission channels such as email.

3.3   Information Extraction Algorithm

The receiver shares the same text-image matching model, the image database and k with the sender. After receiving a text and n images, three steps are performed to extract the secret message.

Step 1: Image sorting. The relevancies between the received text and all the images in the image database are calculated with the text-image matching model. Then all the images are sorted on the basis of the relevancy in descending order.

Step 2: Image matching. The received images are compared to the sorted images one by one. When a received image matches the sorted images, the sort index is recorded and this image is taken from the sorted images. The sort index starts from zero. In this way, n decimal integers can be obtained.

Step 3: Secret message recovery. After converting each decimal number to a binary sequence, all the binary sequences are spliced together into a bit stream. Finally, the secret message is recovered by converting the bit stream.

4   Experiments and Analysis

The performance of the proposed multi-modal method can be evaluated from three aspects, *i.e.*, text-image relevancy, hiding capacity and resistance to the steganalysis methods.

4.1   Text-Image Relevancy

Since side channel steganalysis can detect steganography based on the correlation of the images [8], the relevancies among the text and the companied images will influence the security of the proposed methods. The higher the relevancy is, the more secure the steganography method is.

In order to evaluate the relevancy between the texts and images, we use the test dataset in [6]. The test dataset contains 100 texts and 1,000 images selected from the Twitter100k dataset [7]. All the texts and images are labeled by hand with several tags such as people, clothes and flower.

In [6], the relevancy between the two modalities is measured by matching the tags. If a text and an image have a common tag, the image is regarded as

Secret Message: *China*
It's always good to see your friends from sLOVEenija in Miami !!! #welcomeToMiami

Secret Message : *Japan*
Gorgeous scenery. Beautiful weather. FREEDOM!

Fig. 3. Examples of the proposed multi-modal steganography. Each text is companied with 8 images and each image represents 5 bits.

relevant to the text. Given a text, matching rate@K is defined as the proportion of the relevant images in the top K similar images computed by the text-image matching model. Average matching rate@K is the mean of the matching rate@K of all the texts in text database. When each image represents k bits, the candidate pool contains $K = 2^k$ images. Table 1 presents the average matching rates at different values of k when using baseline VSE model and our text-image matching model improved with adversarial learning and circle loss.

Table 1. The average matching rates at different values of k when using baseline VSE model and our text-image matching model improved with adversarial learning and circle loss.

k		1	2	3	4	5	6
K		2	4	8	16	32	64
Baseline model		0.86	0.84	0.83	0.81	0.79	0.78
Ours	Baseline model + adversarial learning	0.88	0.87	0.86	0.84	0.82	0.80
	Baseline model + circle loss	0.91	0.90	0.87	0.86	0.84	0.81
	Baseline model + adversarial learning + circle loss	0.92	0.91	0.88	0.87	0.85	0.82

As shown in Table 1, both adversarial learning and circle loss are beneficial to the matching rate. The improvement of the average matching rate is nearly 2% when adversarial loss is employed and the benefit of circle loss is about 4%. When adversarial learning and circle loss are utilized simultaneously, the average matching rate increases by 5% approximately. The experiment results verify that adversarial learning and circle loss are useful for text-image matching.

Some examples of the proposed method are given in Fig. 3. Each secret message is an English word of five characters, which can be converted into 40-bit stream. The values of n and k are set to 8 and 5, respectively. Therefore, each text is attached with 8 images and each image represents 5-bits binary sequence.

4.2 Hiding Capacity

The hiding capacity of the proposed method is the same as the multi-modal steganography method proposed in [6]. The hiding capacity is proportional to the number of attached images with a text and the bits represented by each image. When the text is companied with n images and each image represents k-bits binary sequence, the hiding capacity of the proposed method is $k \times n$ bits.

According to the experimental results in [6], the matching rate decreases when the image database is shortened, as shown in Table 2. Therefore, n and k are limited by the size of the image database. However, the image database can be enlarged with the public Twitter100k dataset [7] without much effort. Besides, n is also limited by the restriction of the transmission network, which is very loose in many transmission channels.

Table 2. The average matching rates at different values of k when the size of the image database is 500 and 1,000, respectively.

k	1	2	3	4	5	6
K	2	4	8	16	32	64
Size = 500	0.83	0.81	0.80	0.79	0.77	0.74
Size = 1,000	0.86	0.84	0.83	0.81	0.79	0.78

4.3 Resistance to the Steganalysis Methods

The resistance to the steganalysis methods is crucial for a steganography method. However, a variety of the mainstream steganography methods can be detected by the existing steganalysis methods based on the distortions between the covers after embedding and the original covers [13,14,16,17]. Different from these steganography methods, the proposed multi-modal steganography hides secret message in the relevancy between the text and the image without modifying or generating the cover. All the covers keep natural and original. Therefore, existing steganalysis methods based on the distortions of the covers after embedding are unable to detect our method. Meanwhile, since the selected text and images are highly relevant, the proposed steganography method can also escape from the side channel steganalysis based on the correlation of the images [8].

5   Conclusion

This paper proposes a multi-modal steganography method based on a refined text-image matching algorithm. The text-image matching model is improved by using adversarial learning and circle loss. Experimental results shows that the performance of our model for text-image matching is superior to the baseline model. Due to the high relevancy between the selected text and images, the proposed steganography method can escape from the side channel steganalysis based on the correlation of the images. Besides, since the covers transmitted in the proposed steganography method are not modified or synthesized, existing steganalysis methods based on the distortions between the covers after embedding and the original covers are unlikely to detect our method.

References

1. Cho, K., et al.: Learning phrase representations using RNN encoder-decoder for statistical machine translation. arXiv preprint arXiv:1406.1078 (2014)
2. Du, Y., Yin, Z., Zhang, X.: Improved lossless data hiding for JPEG images based on histogram modification. Comput. Mater. Continua **55**(3), 495–507 (2018)
3. Duan, X., Song, H., Qin, C., Khan, M.K.: Coverless steganography for digital images based on a generative model. Comput. Mater. Continua **55**(3), 483–493 (2018)
4. Faghri, F., Fleet, D.J., Kiros, J.R., Fidler, S.: VSE++: improving visual-semantic embeddings with hard negatives. arXiv preprint arXiv:1707.05612 (2017)
5. Fridrich, J., Kodovsky, J.: Rich models for steganalysis of digital images. IEEE Trans. Inf. Forensics Secur. **7**(3), 868–882 (2012)
6. Hu, Y., Li, H., Song, J., Huang, Y.: MM-stega: multi-modal steganography based on text-image matching. In: Sun, X., Wang, J., Bertino, E. (eds.) ICAIS 2020. CCIS, vol. 1254, pp. 313–325. Springer, Singapore (2020). https://doi.org/10.1007/978-981-15-8101-4_29
7. Hu, Y., Zheng, L., Yang, Y., Huang, Y.: Twitter100k: a real-world dataset for weakly supervised cross-media retrieval. IEEE Trans. Multimed. **20**(4), 927–938 (2018)
8. Li, L., Zhang, W., Chen, K., Zha, H., Yu, N.: Side channel steganalysis: when behavior is considered in steganographer detection. Multimed. Tools Appl. **78**(7), 8041–8055 (2019)
9. Liu, R., Zhao, Y., Wei, S., Zheng, L., Yang, Y.: Modality-invariant image-text embedding for image-sentence matching. ACM Trans. Multimed. Comput. Commun. Appl. **15**(1), 27 (2019)
10. Russakovsky, O., et al.: Imagenet large scale visual recognition challenge. Int. J. Comput. Vis. **115**(3), 211–252 (2015)
11. Simonyan, K., Zisserman, A.: Very deep convolutional networks for large-scale image recognition. arXiv preprint arXiv:1409.1556 (2014)
12. Sun, Y., et al.: Circle loss: a unified perspective of pair similarity optimization. In: IEEE Conference on Computer Vision and Pattern Recognition, pp. 6398–6407 (2020)
13. Wu, S., Zhong, S., Liu, Y.: Deep residual learning for image steganalysis. Multimed. Tools Appl. **77**(9), 10437–10453 (2017). https://doi.org/10.1007/s11042-017-4440-4
14. Xu, G., Wu, H., Shi, Y.: Structural design of convolutional neural networks for steganalysis. IEEE Signal Process. Lett. **23**(5), 708–712 (2016)
15. Yang, Z., Guo, X., Chen, Z., Huang, Y., Zhang, Y.: RNN-stega: linguistic steganography based on recurrent neural networks. IEEE Trans. Inf. Forensics Secur. **14**(5), 1280–1295 (2018)
16. Yang, Z., Huang, Y., Zhang, Y.: A fast and efficient text steganalysis method. IEEE Signal Process. Lett. **26**(4), 627–631 (2019)

17. Yang, Z., Wang, K., Li, J., Huang, Y., Zhang, Y.: TS-RNN: text steganalysis based on recurrent neural networks. IEEE Signal Process. Lett. **26**(12), 1743–1747 (2019). https://doi.org/10.1109/LSP.2019.2920452
18. Yang, Z., Zhang, S., Hu, Y., Hu, Z., Huang, Y.: VAE-stega: linguistic steganography based on variational auto-encoder. IEEE Trans. Inf. Forensics Secur. **16**, 880–895 (2020)
19. Zhang, Y., Ye, D., Gan, J., Li, Z., Cheng, Q.: An image steganography algorithm based on quantization index modulation resisting scaling attacks and statistical detection. Comput. Mater. Continua **56**(1), 151–167 (2018)
20. Zheng, S., Wang, L., Ling, B., Hu, D.: Coverless information hiding based on robust image hashing. In: Huang, D.-S., Hussain, A., Han, K., Gromiha, M.M. (eds.) ICIC 2017. LNCS (LNAI), vol. 10363, pp. 536–547. Springer, Cham (2017). https://doi.org/10.1007/978-3-319-63315-2_47
21. Zhou, Z., Mu, Y., Wu, Q.M.J.: Coverless image steganography using partial-duplicate image retrieval. Soft. Comput. **23**(13), 4927–4938 (2018). https://doi.org/10.1007/s00500-018-3151-8
22. Zhou, Z., Sun, H., Harit, R., Chen, X., Sun, X.: Coverless image steganography without embedding. In: Huang, Z., Sun, X., Luo, J., Wang, J. (eds.) ICCCS 2015. LNCS, vol. 9483, pp. 123–132. Springer, Cham (2015). https://doi.org/10.1007/978-3-319-27051-7_11
23. Zhou, Z., Wu, Q.J., Yang, C.N., Sun, X., Pan, Z.: Coverless image steganography using histograms of oriented gradients-based hashing algorithm. J. Internet Technol. **18**(5), 1177–1184 (2017)

Constructing Immune Cover for Secure Steganography Based on an Artificial Immune System Approach

Hongxia Wang[1], Zhilong Chen[2], and Peisong He[1(✉)]

[1] School of Cyber Science and Engineering, Sichuan University,
Chengdu 610065, People's Republic of China
gokeyhps@scu.edu.cn
[2] School of Information Science and Technology, Southwest Jiaotong
University, Chengdu 611756, People's Republic of China

Abstract. Artificial Immune Systems (AIS) are a class of computationally intelligent systems inspired by the principles and processes of the Biological Immune System (BIS), and has been applied to many fields successfully by exploit its characteristics of self-learning and self-organizing. In this paper, we open a new field for applications of AIS, namely, immune cover construction for steganography application. We proposed an AIS-based steganography framework to improve the security of steganography by constructing immune cover image in spatial domain. Texture complexity is a major factor in resisting steganalysis in images, so the proposed framework is designed by immune processing to adaptively accentuate the texture region as well as maintain the original characteristics of images, and then obtain more suitable and secure immune cover image for steganography. This approach allows to construct more setganalysis-secure data embedding using standard steganography algorithms. Compared with state-of-the-art methods, the proposed method has an improved ability to resist steganalysis.

Keywords: Image steganography · Artificial Immune System · Clonal selection algorithm · Setganalysis

1 Introduction

Steganography is the practice of concealing a secret message ("payload") within another non-secret media ("cover") in the most inconspicuous manner possible. In order to achieve good concealment performance, researchers hope to make least modification cost while ensuring the maximum embedding capacity [1]. Consequently, they often design a series of strategies to meet this requirement. For example, the prevailing steganographic schemes in spatial domain, such as HUGO [2], S-UNIWARD [3], HILL [4], MDS-UNIWARD [5], are all focused on the design of effective distortion function. In addition, the JPEG steganographic schemes by defining cost functions is also proposed [6]. The minimal embedding distortion steganography schemes embed data in less detectable regions and achieving higher security. However, the usage of heuristically defined distortions is inevitably limited when it is used in different scenarios, different embedding domains and different covers.

© Springer Nature Switzerland AG 2021
X. Zhao et al. (Eds.): IWDW 2020, LNCS 12617, pp. 53–67, 2021.
https://doi.org/10.1007/978-3-030-69449-4_5

In recent years, with the development of artificial intelligence, the deep learning-based steganography has paid more attentions. However, most of the existing schemes are limited to steganography modes based on GAN(Generative Adversarial Networks) [7, 8] and CNN (Convolutional Neural Networks) [9, 10], and the security needs to be further improved. Therefore, researchers look forward to combining more deep learning models or other intelligent systems to expand the intelligent model of steganography. Artificial Immune System (AIS) as an important branch in the field of artificial intelligence, has flourished in recent years [11]. Artificial immune system inspired by the concepts of the human immune system to solve computational problems. Many studies have shown that AIS is an efficient technique for optimization problems. There are different AIS algorithms, such as clonal selection algorithm, negative selection algorithm, immune network algorithm, and immune genetic algorithm, etc. These algorithms have been studied mainly from the perspective of learning and memory mechanisms of the immune system [12]. Perez *et al.* [13] proposed a universal steganography detector based on an artificial immune system for JPEG images, which can detect JPEG images modified with three well-known steganographic tools: F5, Outguess or Steghide. El-Emama *et al.* [14] presented a novel image steganography using a new intelligent technique, i.e., estimating the number of bits to be hidden at each pixel with an adaptive genetic algorithm. In this paper, we are devoted to designing a new steganography framework based on an artificial immune system approach. Before data embedding, we first carry on the immune processing to the original cover image, and obtain the immune cover image, and then embed the message into the immune cover image, which makes the embedding traces are less detectable. Therefore, the security performance are improved.

The rest of this paper is organized as follows. In Sect. 2, we describe the proposed steganography framework based on AIS. The immune cover construction method based on AIS is presented in Sect. 3, which is followed by the experimental results and performance comparison in Sect. 4. Finally, conclusions are drawn in Sect. 5.

2   Proposed Steganography Framework Based on Artificial Immune System

Artificial immune system is an intelligent system that can learn from the mechanism and characteristics of Biological Immune System (BIS) to prevent the external intrusion. This is very similar to the security requirement of steganography to resist the statistical detection. Inspired by this principle, we can consider to utilize the intelligent information processing ability of AIS to design a secure steganography framework. This section summarizes the AIS-based steganography framework via immune cover as shown in Fig. 1. First, the original cover image is optimized by an AIS approach to obtain a new cover. This new cover via immune processing is called immune cover image for information embedding. Moreover, the security of embedded information in this immune cover is higher than that of the original cover image, then we use the existing standard steganography algorithm to embed the secret message into the immune cover image to obtain the stego-image.

Fig. 1. Steganography framework based on artificial immune system.

In order to make AIS-based steganography framework can well simulate the mechanism of immune system, we need construct the mapping relationship of basic components between biological immune system, artificial immune system and AIS-based steganography framework. The immune system detects the foreign substances which enters into or contacts with the body. Biological immune system can effectively protect the body against pathogens and harmful microorganisms, while the artificial immune system is a kind of computational intelligence which adopts the immune behavior as biological immune system. The artificial immune algorithms typically exploit the immune system's characteristics of self-learning, self-organization, self-adaptive and immune memory to solve effectively complex problems. Inspired by the principle of biological immunity, we can regard the statistical changes caused by data embedding as the pathogens invading the body. Based on the principle of artificial immune, the design of secure steganography method can be seen as how to eliminate these "pathogens" as possible, namely, eliminating these statistical changes. Therefore, we can establish an AIS-based steganography model to make the statistical changes as small as possible. The mapping relationship between the basic components of biological immune, artificial immune and AIS-based steganography model is listed in Table 1.

Table 1. Mapping relationship of basic components.

BIS	AIS	AIS-based steganography framework
Antigen	Objective and constraints	Statistical changes caused by steganography
Antibody	Candidate solutions of problem	Solutions to eliminate statistical changes
Fitness	Ability of pattern matching	Ability of eliminating statistical changes
Immune response	Solution of problem	Elimination of statistical changes

In Table 1, the antigens in the biological immune system represent all kinds of external pathogens that enter the body and start a process that can cause disease. The body then usually produces antibodies to fight the antigens. For the steganography, data embedding will change the statistical characteristics of the cover image, which is similar to the impact of antigens on the body. Therefore, we can correspond the antigen to the statistical changes caused by steganography. The different steganography algorithms have different statistical changes to the cover image, so the antibodies should be mapped to solutions to eliminate these statistical changes from a security point of view since the most secure steganographic schemes require minimal distortion embedding. The fitness is initially determined by the stimulation response of a particular antibody molecule to the currently presented antigenic pattern in BIS, while the fitness concept is essential in order to evaluate the pattern matching ability in AIS. Consequently, the fitness is used to measure the extent of eliminating statistical changes in the AIS-based steganography framework, and the most secure steganographic scheme can be obtained according to the fitness value.

3 Immune Cover Construction Based on Artificial Immune System

Recent years, many content-adaptive steganographic schemes with minimal distortion embedding tend to embed the secret message into highly textured regions of images, where the embedding traces are less detectable [15]. However, not every image has rich texture for steganography. Moreover, the modified pixels caused by steganography are concentrated in the texture rich regions, so the embedding changes are also concentrated in the texture regions. For this reason, we consider to accentuate the texture of the cover image for immune processing in this paper. We define the cover image after immune processing to be immune cover image. In order to obtain a more suitable and secure cover image for steganography, we use the clonal selection algorithm to optimize the parameters of immune processing. When the parameters are optimal, the most suitable and secure immune cover image for steganography can be obtained. We then embed the secret message into this optimal immune cover image using an existing standard steganographic scheme. Thus the embedded message is more difficult to be detected by steganalysis techniques.

3.1 Immune Processing of Original Cover Image

The second-order derivative can be used to detect edges in an image. Since an image is actually a two-dimensional signal, we would need to take the derivative in both dimensions. Here, the Laplacian operator comes handy. The Laplacian operator of f is defined by [16]

$$\nabla^2 f = \frac{\partial^2 f}{\partial x^2} + \frac{\partial^2 f}{\partial y^2} \tag{1}$$

where ∇^2 denotes the Laplacian operator. In fact, since the Laplacian uses the gradient of images, it uses neighborhood pixels to improve local contrast and accentuate the texture of images. For a digital image, we use the following filter template shown in Fig. 2 to accentuate the texture of images.

Fig. 2. Filter template for enhancing the texture of images.

The 8-bit grayscale image and its texture accentuated version (with dimensions of $n_1 \times n_2$) are denoted by $X = (x_{ij})^{n_1 \times n_2}$ and $Y = (y_{ij})^{n_1 \times n_2}$, $x_{ij}, y_{ij} \in \{0, 1, \ldots, 255\}$. The accentuated pixel y_{ij} of pixel x_{ij} can be calculated by

$$y_{ij} = x_{ij} + \alpha z'_{ij}, \quad \alpha > 0 \tag{2}$$

and

$$z'_{ij} = \begin{cases} z_{ij}, & z_{ij} < T \\ T, & other \end{cases} \tag{3}$$

where α denotes the intensity factor that controls the extent of image texture, $T > 0$ is a threshold. The purpose of setting T is to avoid greatly reducing image quality due to over accentuation to some pixels. In addition, z_{ij} can be computed as

$$z_{ij} = 8x_{ij} - x_{i+1,j} - x_{i-1,j} - x_{i,j+1} - x_{i,j-1} - x_{i-1,j-1} - x_{i-1,j+1} - x_{i+1,j-1} - x_{i+1,j+1} \tag{4}$$

The larger the value of intensity factor α, the greater the extent of texture accentuation. We will use the clonal selection algorithm to optimize α to get the optimal accentuated image as immune cover image. The optimization process of intensity factor α using clonal selection algorithm is immune processing to the original cover images. After immune processing, we get the immune cover image for message embedding. Different values of α will obtain different immune cover image, and the purpose of immune processing is to get immune cover images more suitable and secure for steganography, and for this reason, we use intensity factor α as an antibody in the clonal selection algorithm.

In order to verify whether the intensity factor α is reasonable as an antibody and whether it can improve the security of steganography algorithm, we set $T = 8$ and execute four times experiments under $\alpha = 0, 0.3, 0.6, 0.9$, respectively. All experiments are conducted on 2000 grayscale images of size 512×512 that are downloaded from BOSSbase ver. 1.01 [17]. We use HUGO algorithm [2] with a payload 0.4 bpp (bit per pixel) to embed the secret messages, and the performances are evaluated with steganalyzers using the 34671-D SRM (Spatial Rich Models) feature set [18] with ensemble classifiers [19]. We randomly selected 1000 images for training and used the remaining 1000 images for testing. The security performance is expressed as the average detection error rate P_E defined as [2]. We compare the security performance under different intensity factor α when resisting SRM steganalysis in Fig. 3, from which we can see that generally the larger α is and the greater the detection error rate P_E is. That is, with the increate of α, the security performance is improved. Note that $\alpha = 0$, the steganography will be HUGO algorithm. So the immune processing can improve the security performance of steganography algorithm. However, the larger α will result in the lower quality of immune cover image. If we embed the message into the immune cover image with lower quality, the quality of stego-image will be also low. Therefore, we need to optimize the intensity factor α as an antibody using adaptive immune clonal selection algorithm, namely, the optimization process can get more suitable and secure immune cover image for steganography.

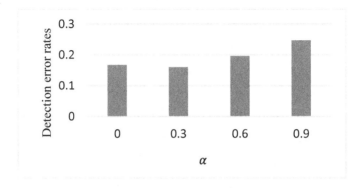

Fig. 3. Detection error rates P_E under different intensity factor α.

3.2   Optimization Process of Immune Cover Image

As mentioned previously, the intensity factor α will act as the antibody. In fact, the optimization of immune cover image is also the optimization of antibody α value. In this paper, we apply clonal selection algorithm to optimize antibody α value. Firstly, the expression forms of antibody need be determined. In artificial immune system, there are many kinds of expression forms such as binary code and real number code. Here, binary form is selected to encode antibody α value, and the length of coding is calculated by

$$L = \log_2 \left(1 + \frac{v_{\max} - v_{\min}}{\mu} \right) \tag{5}$$

where μ is the encoding precision, and v_{min}, v_{max} represent the minimum and maximum values of the antibody α, respectively. The corresponding real value of antibody α can be obtained by decoding as follows:

$$v = v_{min} + \mu \sum_{i=1}^{L} \beta_i 2^{i-1}, \ \beta_i \in \{1, 0\}, \ i = 1, 2, \ldots, L \tag{6}$$

where β_i denotes the i-th component of antibody coding with binary form. In the process of optimizing antibody and improving antibody fitness, the parent antibody is cloned and mutated to produce offspring antibody. The expression of mutation operator y_i is given by

$$y_i = x_i - 2x_i c_i + c_i, \ 1 \leq i \leq L \tag{7}$$

and

$$c_i = \begin{cases} 1 & r_i \leq P_{mu} \\ 0 & other \end{cases} \tag{8}$$

where r_i is a random number and $r_i \in [0, 1]$, and P_{mu} is the mutation rate. The mutation is implemented by randomly changing one of the bits in a given string and inversion is implemented by inverting a randomly chosen segment of the string. Here, if $c_i = 0$, the antibody coding bit will remain unchanged, else, $c_i = 1$, the antibody coding bit will be reversed. That is, 0 is changed to 1, and 1 is changed to 0, thus the antibody mutation is implemented.

Let n be the number of clones, and the clonal operator may be formulated as

$$n = 10 \times \frac{\eta - \eta_{\min}}{\eta_{\max} - \eta_{\min}} \tag{9}$$

in which η denotes the fitness of an antibody, and $\eta_{\max}$, $\eta_{\min}$ correspond to the maximum and minimum fitness values in the current antibody population respectively. A number of high affinity antibodies are selected and reproduced (clonal) according to their fitness. That is, the higher the fitness of antibody, the higher the clonal rate. Those antibodies whose fitness is less than a pre-specified threshold are eliminated (clonal suppression) with an elimination rate.

The number of clones produced for each antibody is proportional to the measure of its fitness with a given antigenic pattern. The clones generated undergo mutation inversely proportional to their antigenic affinity. That is, the higher the fitness, the lower the mutation rate. Since the AIS-based steganography is expected to obtain a more suitable cover image for data embedding, the fitness of antibody should be related to the detectability of the stego-image. The detection performance is mainly determined by the embedding changes of the image. The smaller the embedding changes, the greater the detection error rate. Namely, the security performance is higher. In this paper, we compute the Euclidean distance between the statistical features of the stego-image and the immune cover image as the fitness of antibodies. So the fitness can be calculated as

$$\eta = \frac{1}{\sqrt{\sum_{i=1}^{N} (F_i^c - F_i^s)^2}} \tag{10}$$

where N is the dimension of statistical features such as SRM, and F_i^c, F_i^s are the i-th dimensional feature component of immune cover image and stego-image, respectively.

Based on this fitness, we apply the clonal selection algorithm to optimize the intensity factor α as the antibody to obtain the best one. That is, the most suitable and secure immune cover image can be obtained for stenography. The block diagram of optimization process of intensity factor α based on clonal selection algorithm is shown in Fig. 4. The following describes the steps of optimization process.

1) Initialization: the intensity factor α of the texture accentuation acts as the antibody, and the population size is the number of antibodies that works in each generation. According to the population size of antibodies, the initial population of antibodies with binary form is generated randomly based on Eq. (5).

2) Calculation of fitness: according to the intensity factor α, the texture accentuated image as an immune cover image is obtained. Then, we calculate fitness value after embedding secret message in an immune cover image based on Eq. (10).

3) Selection of the best α based on fitness and clonal rate: the clonal rate is between 0 and 1 that is used to get the number of clones an antibody. According to the antibody fitness, the optimal antibodies are selected according to the clone rate and cloned to form a clone pool.

4) Antibody mutation based on fitness and mutation rate: the cloned antibody will be mutated. Mutation rate is between 0 and 1 that is the probability of a given feature will be mutated, and the mutated antibodies and their number can be obtained according to mutation rate and Eq. (7). This mutated antibodies are different from their parents.

5) Replace parent antibody: calculate the fitness after antibody mutation in the clone pool. If the fitness of the mutated antibody is higher than that of the parent antibody, the parent antibody will be replaced.

6) Replace the antibody whose fitness is the worst: the antibodies with the lowest fitness are eliminated according to the elimination rate. At the same time, new antibodies are generated randomly to replace the eliminated individuals to maintain the diversity of antibody set and avoid falling into local optimization.

7) After completing one time iteration, a new population is generated. Let the pre-specified iteration time be the stop condition. If the iteration time meets the stop condition, the best antibody is output. Namely, the best intensity factor α of the texture accentuation can be obtained.

Consequently, we use this best intensity factor α to get the most suitable and secure immune cover image for data embedding. Finally, we use the existing standard steganography algorithms to embed message into the obtained immune cover image, and get the stego-image.

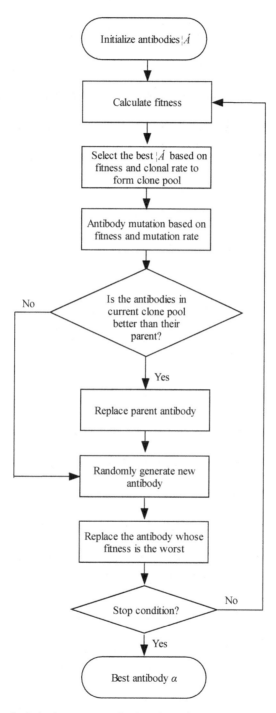

Fig. 4. Optimization process of α based on clonal selection algorithm.

4   Experimental Results and Analysis

4.1   Experiment Setups

All experiments are carried out on the image database BOSSbase ver. 1.01 [17], which contains 10,000 gray-scale images of size 512×512. Before data embedding, we need to perform immune processing on the original cover image to get the immune cover image. For purpose of immune processing of the original cover image, we use clonal selection algorithm to optimize the original cover image. The intensity factor α that controls the extent of image texture will act as the antibody during optimizing the original cover image. The parameters of clonal selection algorithm are set in Table 2.

Table 2. Parameters of clone selection algorithm.

Parameters	Value
Initial population size of antibodies	30
Iteration times	20
Clonal rate	0.3
Mutation rate	0.2
Elimination rate	0.2
Intensity factor α as the antibody	[0, 0.7]
Coding precision	0.01

After immune processing on the original cover image, we get the immune cover image for embedding the secret message. In our experiments, HUGO as a basic steganography method is employed for data embedding. In HUGO algorithm, the parameters σ, γ can be tuned in order to minimize the detectability. In our experiments, we set σ, γ to be the optimal values, namely, $\sigma = 10$, $\gamma = 4$. We use the simulated optimal embedding as the default for HUGO algorithm and the performances are evaluated with steganalyzers using the 34671-D SRM feature set [18] and maxSRM feature set [20], where the Fisher Linear Discriminants (FLD) ensemble classifier [19] with default settings is used to train the binary classifiers. We randomly selected 5000 images for training and used the remaining 5000 images for testing. The classification error probability P_E of FLD ensemble classifier is reported by the mean value of the ensemble's testing errors based on ten times of randomly testing and different embedding payloads.

4.2   Imperceptibility

Five gray images of size 512×512, Lena, Baboon, House, Lake and Pepper shown in Fig. 5, which contain landscape, animal, building and people are used as the original cover images in the experiment. According to our method, we get the immune cover images shown in Fig. 6. Then we embed the secret message into the obtained immune cover images when the payload rate is 0.4 bpp, and get the stego-images shown in Fig. 7. From Fig. 5, 6 and 7, we see that there are no obvious visual abnormality between original cover images, immune cover images and stego-images.

Fig. 5. Original cover images. (a) Lena, (b) Baboon, (c) House, (d) Lake, (e) Peppers.

Fig. 6. Immune cover images. (a) Lena, (b) Baboon, (c) House, (d) Lake, (e) Peppers.

Fig. 7. Stego-images. (a) Lena, (b) Baboon, (c) House, (d) Lake, (e) Peppers.

Table 3 and Table 4 list the PSNR (Peak Signal to Noise Ratio) values between stego-images and original cover images, and between stego-images and immune cover images, at different embedding payloads from 0.1 to 0.5 bpp. It can be seen that all of PSNR values are larger than 38 dB in Table 3, and larger than 51 dB in Table 4. Therefore, our scheme has a good imperceptibility.

Table 3. PSNR (dB) between stego-images and original cover images under different embedding payload (bpp).

Image	Payload				
	0.1	0.2	0.3	0.4	0.5
Lena	40.441	40.377	40.315	40.244	40.179
Baboon	39.173	39.129	39.082	39.034	38.982
House	46.568	46.370	46.175	45.957	45.741
Lake	39.645	39.603	39.554	39.511	39.459
Peppers	38.864	38.820	38.770	38.717	38.663

Table 4. PSNR (dB) between stego-images and immune cover images under different embedding payload (bpp).

Image	Payload				
	0.1	0.2	0.3	0.4	0.5
Lena	59.961	56.630	54.732	53.243	52.117
Baboon	60.742	57.168	55.014	53.510	52.255
House	60.757	57.417	55.519	54.087	52.909
Lake	60.642	57.235	55.144	53.644	52.507
Peppers	59.900	56.538	54.550	53.031	51.847

4.3 Non-detectability

We adopt two popular feature sets to detect the existence of secret message in our stego-image, including 34671-D SRM feature set and maxSRM feature set. The FLD ensemble classifier is used for training and classification [19]. For the image database containing 10,000 gray-scale images, half of the images and half of the stego-images are used for training, while the rest are used for testing, where the stego-images are generated using the HUGO at a specific payload from 0.1 to 0.5 bpp. Such process is repeated ten times to obtain an average detection error rate P_E. We compare the performance of the non-detectability of our stego-images with other steganography schemes under SRM feature set and maxSRM feature set, respectively (basic steganography HUGO). And the results are illustrated in Fig. 8. Let's denote our proposed framework with the use of the basic steganography algorithm HUGO as the Immune-HUGO. It can be seen from Fig. 8 that the P_E values of our scheme is larger than that of HUGO, S-UNIWARD and MDS-UNIWARD schemes under the five payload rates, which indicates that the non-detectability of our scheme against SRM and maxSRM statistical analysis is superior to that of HUGO, S-UNIWARD and MDS-UNIWARD. It is verified that immune processing can improve the security of steganography algorithm to resist statistical detection. Even HUGO algorithm with lower security can also be improved by combining with artificial immune algorithm.

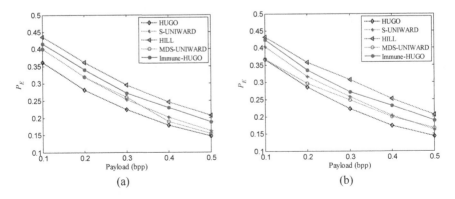

Fig. 8. Comparison results of non-detectability under basic steganography HUGO. (a) SRM feature, (b) maxSRM feature.

However, we can also see from Fig. 8 that the security of our scheme is still lower than that of HILL scheme. The reason is that our scheme is based on HUGO, and the security of HUGO algorithm is far lower than that of HILL, so the security of our scheme is higher than that of HUGO, but lower than HILL. What happens if we replace the basic steganography algorithm HUGO with another, more secure steganography algorithm? Let's do an extending experiments by replacing the basic steganography algorithm HUGO with S-UNIWARD, and further study the non-detectability of our scheme. After replacing HUGO with S-UNIWARD, we compare the non-detectability of our stego-images with other steganography schemes in Fig. 9. Let's denote our proposed framework with the use of basic steganography algorithm S-UNIWARD as the Immune-UNIWARD. It can be seen that our scheme clearly outperforms HUGO, S-UNIWARD, MDS-UNIWARD and HILL at different embedding payloads from 0.1 to 0.5 bpp. Extending experiments show that the proposed methods (Immune-UNIWARD) can achieve higher level of security than the original methods.

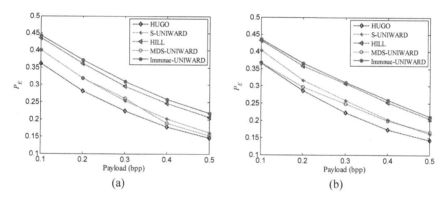

Fig. 9. Comparison results of non-detectability under basic steganography S-UNIWARD. (a) SRM feature, (b) maxSRM feature.

5 Conclusions

In this paper, we propose a new AIS-based steganography framework by constructing the immune cover image. Different from the previous schemes, the immune cover image is firstly constructed by the optimization of clonal selection algorithm, and then the secret message is embedded into the immune cover image by the existing steganography such as HUGO and S-UNIWARD, thereby effectively defending the steganalysis attack. Compared with state-of-the-art methods, our method has an improved security to resist the SRM and maxSRM detection. The proposed method provides a new solution for designing the intelligent steganography scheme based on artificial immune system. Also, the strategy of constructing the immune cover image could be extended with more advanced steganographic algorithms, such as HILL. In the future, the AIS-based steganography framework could be extended to the JPEG domain.

Acknowledgments. This work was supported by National Natural Science Foundation of China (61972269, 61902263), China Postdoctoral Science Foundation (2020M673276), and the Fundamental Research Funds for the Central Universities (YJ201881, 2020SCU12066).

References

1. Sharifzadeh, M., Aloraini, M., Schonfeld, D.: Adaptive batch size image merging steganography and quantized Gaussian image steganography. IEEE Trans. Inf. Forensics Secur. **15**(7), 867–879 (2020). https://doi.org/10.1109/tifs.2019.2929441
2. Pevný, T., Filler, T., Bas, P.: Using high-dimensional image models to perform highly undetectable steganography. In: Böhme, R., Fong, P.W.L., Safavi-Naini, R. (eds.) IH 2010. LNCS, vol. 6387, pp. 161–177. Springer, Heidelberg (2010). https://doi.org/10.1007/978-3-642-16435-4_13
3. Holub, V., Fridrich, J., Denemark, T.: Universal distortion function for steganography in an arbitrary domain. EURASIP J. Inf. Secur. **2014**(1), 1–13 (2014). https://doi.org/10.1186/1687-417x-2014-1
4. Li, B., Wang, M., Huang, J., Li, X.: A new cost function for spatial image steganography. In: IEEE International Conference on Image Processing, pp. 4206–4210 (2014). https://doi.org/10.1109/icip.2014.7025854
5. Bian, Y., Tang, G., Gao, Z., Wang, S.: A distortion cost modification strategy for adaptive pentary steganography. Multimed. Tools Appl. **76**(20), 20643–20662 (2017). https://doi.org/10.1007/s11042-016-3986-x
6. Chen, K., Zhou, H., Zhou, W., Zhang, W., Yu, N.: Defining cost functions for adaptive JPEG steganography at the microscale. IEEE Trans. Inf. Forensics Secur. **14**(4), 1052–1066 (2019). https://doi.org/10.1109/tifs.2018.2869353
7. Volkhonskiy, D., Borisenko, B., Burnaev, E.: Generative adversarial networks for image steganography. In: 5th International Conference on Learning Representations (ICLR), pp.1–8 (2017)
8. Hayes, J., Danezis, G.: Generating steganographic images via adversarial training. In: Advances in Neural Information Processing Systems (NIPS), pp. 1951–1960 (2017)
9. Tang, W., Li, B., Tan, S., Barni, M., Huang, J.: CNN-based adversarial embedding for image steganography. IEEE Trans. Inf. Forensics Secur. **14**(8), 2074–2087 (2019). https://doi.org/10.1109/tifs.2019.2891237
10. Wu, J., Chen, B., Luo, W., Fang, Y.: Audio steganography based on iterative adversarial attacks against convolutional neural networks. IEEE Trans. Inf. Forensics Secur. **15**(1), 2282–2294 (2020). https://doi.org/10.1109/tifs.2019.2963764
11. Ahmadi, S., Sajedi, H.: Image steganography with artificial immune system. In: Artificial Intelligence and Robotics (IRANOPEN), pp. 45–50 (2017). https://doi.org/10.1109/rios.2017.7956442
12. Bahekar, K., Saurabh, P., Verma, B.: Improving high embedding capacity using artificial immune system: a novel approach for data hiding. In: Kumar, V., Bhatele, M. (eds.) Proceedings of All India Seminar on Biomedical Engineering (AISOBE). Lecture Notes in Bioengineering, pp. 209–219. Springer, India (2013). https://doi.org/10.1007/978-81-322-0970-6_24
13. Pérez, J., Rosales, M., Cortés, N.: Universal steganography detector based on an artificial immune system for JPEG images. In: IEEE TrustCom-BigDataSE-ISPA, pp. 1896–1903 (2016)

14. El-Emama, N., Al-Diabatb, M.: A novel algorithm for colour image steganography using a new intelligent technique based on three phases. Appl. Soft Comput. **37**, 830–846 (2015). https://doi.org/10.1016/j.asoc.2015.08.057

15. Su, W., Ni, J., Li, X., Shi, Y.: A new distortion function design for JPEG steganography using the generalized uniform embedding strategy. IEEE Trans. Circ. Sys. Video Technol. **28**(12), 3545–3549 (2018). https://doi.org/10.1109/tcsvt.2018.2865537

16. Shi, Y., Sun, H.: Image and Video Compression for Multimedia Engineering Fundamentals, Algorithms, and Standards, 3rd edn. Taylor & Francis Group, New York (2019)

17. Bas, P., Filler, T., Pevný, T.: "Break our steganographic system": the ins and outs of organizing BOSS. In: Filler, T., Pevný, T., Craver, S., Ker, A. (eds.) IH 2011. LNCS, vol. 6958, pp. 59–70. Springer, Heidelberg (2011). https://doi.org/10.1007/978-3-642-24178-9_5

18. Fridrich, J., Kodovsky, J.: Rich models for steganalysis of digital images. IEEE Trans. Inf. Forensics Secur. **7**(3), 868–882 (2012). https://doi.org/10.1109/tifs.2012.2190402

19. Kodovsky, J., Fridrich, J., Holub, V.: Ensemble classifiers for steganalysis of digital media. IEEE Trans. Inf. Forensics Secur. **7**(2), 432–444 (2012). https://doi.org/10.1109/tifs.2011.2175919

20. Denemark, T., Sedighi, V., Holub, V., Cogranne, R.: Selection-channel-aware rich model for steganalysis of digital images. In: IEEE International Workshop on Information Forensics and Security (WIFS), pp. 48–53 (2014). https://doi.org/10.1109/wifs.2014.7084302

Generating JPEG Steganographic Adversarial Example via Segmented Adversarial Embedding

Sai Ma[1,2] and Xianfeng Zhao[1,2(✉)]

[1] State Key Laboratory of Information Security,
Institute of Information Engineering, Chinese Academy of Sciences,
Beijing 100195, China
{masai,zhaoxianfeng}@iie.ac.cn
[2] School of Cyber Security, University of Chinese Academy of Sciences,
Beijing 100195, China

Abstract. Nowadays, Convolutional Neural Network (CNN) based steganalytic schemes further improves the detection ability of the steganalyzer comparing with feature based schemes. Besides steganalysis, CNN model can also be used in steganography. Inheriting the mechanism from adversarial attack to CNN model, adversarial embedding is a kind of steganographic scheme that exploits the knowledge of CNN-based steganalyzer. Adversarial embedding can effectively improve security performance of typical adaptive steganographic schemes. In this paper, we propose a novel adversarial embedding scheme for steganography named as Segmented Adversarial Embedding (SAE). The core of SAE is separating the embedding process into several partial embedding processes and performing adversarial embedding in each segment. In each partial embedding process, there is an individual CNN model corresponding to the current embedding stage. In the embedding process, a novel scheme is applied in the cost adjustment. Comparing with the adjustment scheme that utilizes the gradient sign, the new scheme also takes the gradient magnitude into account, which further makes use of the gradient information. Besides the typical implementation of SAE, we also develop a simplified variant with lower complexity. The evaluations on different kinds of steganalyzer prove that SAE is effective to improve the performance of existing steganographic scheme.

Keywords: Steganography · Adversarial example · Deep learning

1 Introduction

Recently, adversarial example technique is utilized in steganography to enhance the security performance. Typical distortion-minimizing steganographic schemes

X. Zhao—This work is supported by the State Grid Corporation of China Project (5 700-202018268A-0-0-00).

[7,8] define the steganographic model heuristically, while the adversarial schemes exploit the knowledge of adversary, which inherit the concept of ASO (Adaptive Steganography by Oracle) [14]. The adversarial schemes are gradient-based, which are derived from FGSM [9]. Except cover enhancing scheme proposed in [21], most of adversarial schemes utilize gradient of the adversary to modulate the embedding cost. Essentially, gradient-based scheme increases the classification loss of the generated stego sample via gradient ascent, so as to confuse the adversary. With the knowledge of adversary model, the adversarial steganographic schemes can improve the performance of existing distortion-minimizing steganographic schemes.

Adversarial steganographic schemes can be categorized into two types, which are feedback and non-feedback. In feedback scheme, amount of adversarial noise is adjusted according to a specified criterion, in order to ensure the generated adversarial stego sample is misclassified to cover by the target adversary. For example, ADV-EMB [19] is a feedback scheme. In the embedding procedure of ADV-EMB, cover is randomly divided into two non-overlapping parts and embedding is performed in two divided steps. In the first step, the subsequence of the message is embedded in the first part of cover with non-adjusted cost. After the first embedding step, the gradient map to the partial embedded cover is calculated to adjust the embedding cost. In the second step, the rest of the message sequence is embedded into the second part of cover with the adjusted cost. In the feedback step, generated candidate adversarial sample is tested by the adversary. Multiple candidates with different amount of adjusted cost are generated. The sample with the smallest amount and misclassified by the adversary is the chosen one. The feedback scheme can be formalized to an optimal problem to generate the adversarial sample that can precisely fool the adversary. [1,2] explore the game theory in adversarial embedding, i.e., the min-max game. In non-feedback scheme, the amount of adversarial noise is fixed, usually, the cost of every cover element is adjusted by the gradient signal. For example, AEN (Adversarial ENhancing) [16] directly exploits the gradient map to adjust the embedding cost. The scheme proposed in [17] applies multiple iteratively trained adversary to update the embedding cost multiple times. Non-feedback scheme has no criterion to get an optimized adversarial stego sample, while directly makes use of the gradient information to adjust cost.

In this paper we propose a non-feedback steganographic adversarial scheme named Segmented Adversarial Embedding (SAE). In SAE, cover is divided into several segments and the embedding is performed on these parts progressively. There are multiple adversaries applied in the embedding, each adversary model is corresponding to each intermediate embedding result. As the cost adjustment scheme used in [19] is not suitable for SAE, SAE applies a novel cost adjusting method, which is derived from the method used in additive side-informed steganography [4]. In former adversarial embedding schemes [2,16,17,19,21], the adversarial models are trained with fully embedded samples. In contrast, SAE utilizes adversarial models trained with partially embedded samples corresponding to each partial embedding process and generate the stego image progressively.

This kind of 'segmented' embedding process adjusts the embedding cost multiple times, which helps to improve the performance for the non-feedback embedding scheme. The main contributions of our work are as follows:

(1) A novel non-feedback adversarial steganographic scheme is proposed. The scheme generates the adversarial stego sample with multiple adversary progressively, which improves the performance against the steganalyzer.
(2) A novel cost adjusting method is used in the proposed scheme, which can make more use of gradient.
(3) Besides the original version of SAE, another simplified version is proposed, which has lower complexity than the original version and better performance in some cases.

The rest of paper is organized as follows: Sect. 2 introduces the notations used in the paper and cost adjustment. The proposed scheme is introduced in Sect. 3. Section 4 is the experiment part. Section 5 is the conclusion.

2   Preliminary

2.1   Notation

Cover image is noted as $\mathbf{C} = (c_{ij})^{H \times W}$. Conventional stego image, which is not generated via adversarial scheme, is noted as $\mathbf{S} = (s_{ij})^{H \times W}$. Adversarial stego image is noted as $\mathbf{Z} = (z_{ij})^{H \times W}$. H is the height, W is the width, and i, j are coordinate of image elements. Furthermore, the $+1$ cost in the position (i, j) is noted as ρ_{ij}^{+}, while -1 cost is noted as ρ_{ij}^{-}. The image set of a specific kind of sample is noted as $\{\cdot\}$, for example, cover set $\{\mathbf{C}\}$ and stego set $\{\mathbf{S}\}$.

Neural network model in this paper is always a binary classifier and its training set is $\{\mathbf{C}, \mathbf{S}\}$ or $\{\mathbf{C}, \mathbf{Z}\}$. The neural network is noted as $F_{\{\mathbf{C}, \mathbf{S}\}}$ or $F_{\mathbf{S}}$ shortly. Loss function is noted as $J(\mathbf{X}, y, F_{\mathbf{S}})$. In the expression, y is the target category. The gradient of $J(\mathbf{X}, y, F_{\mathbf{S}})$ respects to $\mathbf{X}$ is noted as $\mathbf{G} = (g_{ij})^{H \times W} = \nabla_{\mathbf{X}} J(\mathbf{X}, y, F_{\mathbf{S}})$.

2.2   Adversarial Cost Adjustment

In adversarial steganographic scheme, the cost adjustment is a key process. In FGSM, the adversarial sample is generated via (1).

$$\tilde{\mathbf{X}} = \mathbf{X} - sign(\nabla_{\mathbf{X}} J(\mathbf{X}, t)) \tag{1}$$

In (1), $\tilde{\mathbf{X}}$ is the generated sample being classified to the wrong category t, $-sign(\nabla_{\mathbf{X}} J(\mathbf{X}, t))$ is the adversarial noise derived from gradient added to the image $\mathbf{X}$. In adversarial steganographic scheme, modification introduced by embedding can be utilized as adversarial noise. The embedding cost is modulated by gradient (2).

$$\begin{cases} \rho_{ij}^{+} > \rho_{ij}^{-}, & \text{if } g_{ij} > 0; \\ \rho_{ij}^{+} < \rho_{ij}^{-}, & \text{if } g_{ij} < 0 \end{cases} \tag{2}$$

The initial ρ_{ij}^{+} and ρ_{ij}^{-} are equal, they are adjusted according to g_{ij} to follow the direction of adversarial noise.

3 Proposed Scheme

In this section, the proposed SAE is introduced in detail. Concepts of partial embedding and progressive adversarial training are introduced in Sect. 3.1 and 3.2 respectively. The cost adjustment scheme is introduced in Sect. 3.3. In the end, the implementation of SAE is introduced in Sect. 3.4. It is emphasized that SAE contains two stages, which are training stage and generating stage. In the training stage, target steganalyzers (adversaries) are trained with samples. In the generating stage, the adversarial stego samples are generated via target steganalyzers. The image sets used in two stages are isolated.

3.1 Partial Embedding

Partial embedding is an embedding strategy that dividing the cover and message sequence into several non-overlapping segments and performing embedding on these segments separately. In each segment, embedding costs are updated according to the former embedding result. In SMD (Synchronizing Modification Direction) scheme [6,15,20], cover image is divided into several regularly distributed lattices. The embedding is performed on lattices independently, but in each lattice the cost is adjusted according to the intermediate embedding result of former lattices, in order to encourage the same modification clustering. In adversarial scheme ADV-EMB and its iterative version [2], cover is divided into 2 parts pseudo-randomly. The embedding cost of second part is adjusted according to the first part.

In SAE, cover sequence is divided into several parts averagely, so as to message sequence. Unlike [19] and [2], whose separation length is alterable, the separation length in SAE is fixed. The order of sequence is pseudo-random controlled by seed. In this paper, the number of separating parts T is set to 4. Figure 1 illustrates the procedure of partial embedding used in SAE. The message sequence $\mathbf{m}$ is averagely divided into 4 subsequence. These message sequences are noted as $\mathbf{m}_1$, $\mathbf{m}_2$, $\mathbf{m}_3$ and $\mathbf{m}_4$. The message $\mathbf{m}$ is embedded into cover $\mathbf{C}$, the generated stego image is $\mathbf{S}$. Between $\mathbf{C}$ and $\mathbf{S}$, there are 3 intermediate stego images, which are noted as $\mathbf{S}_1$, $\mathbf{S}_2$ and $\mathbf{S}_3$. The final embedding result $\mathbf{S}$ can be also noted as $\mathbf{S}_4$.

3.2 Progressive Adversarial Training

Besides partial embedding, progressive adversarial training is another important concept of SAE. In [19] and [16], the target adversary model is trained from the cover set $\{\mathbf{C}\}$ and the non-adversarial stego set $\{\mathbf{S}\}$. With the consideration of game theory, the iterative adversarial training is taken into account [2,17]. In these iterative schemes, the adversarial stego sets $\{\mathbf{Z}\}$ are taken into the training process of the adversary models. The 'training-generating' procedures performs multiple times in order to improve the security.

Progressive adversarial training is a sort of iterative schemes. In the training stage of SAE, adversarial training and partial embedding of samples are coupled

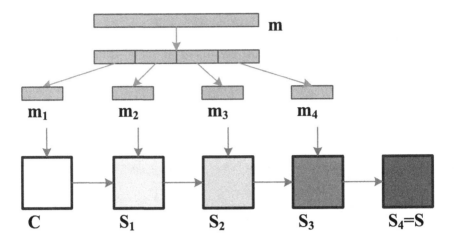

Fig. 1. Partial embedding

together. The training sets used in adversarial training are partial embedded samples. The process of training an adversary model $F_{\mathbf{Z}_i}$ contains 2 steps, which is also illustrated in Fig. 2. Firstly, the image set $\{\mathbf{Z}_i\}$ is generated. For each sample, $\mathbf{Z}_i$ is generated by embedding message $\mathbf{m}_i$ into $\mathbf{Z}_{i-1}$, the embedding cost is adjusted according to the gradient calculated from $F_{\mathbf{Z}_{i-1}}$. Secondly, $F_{\mathbf{Z}_i}$ is trained (fine-tuning) from $F_{\mathbf{Z}_{i-1}}$ with cover set $\mathbf{C}$ and stego set $\mathbf{Z}_i$. The model $F_{\mathbf{Z}_i}$ and set $\mathbf{Z}_i$ are used for the next 'training-generating' round.

3.3   New Cost Adjustment Scheme

A new cost adjustment scheme is applied in SAE. The scheme is inspired by additive side-informed steganography [4]. In FSGM, the adversarial noise is $-sign(\nabla_{\mathbf{X}} J(\mathbf{X}, t))$, which utilizes the sign of the gradient signal. One common cost adjustment scheme used in feedback adversarial scheme is adjusting the cost with constant $\nu > 1$, whose expression is (3).

$$
\rho_{ij}^{+'} = \begin{cases} \rho_{ij}^{+}/\nu, & \text{if } g_{ij} < 0; \\ \rho_{ij}^{+} \cdot \nu, & \text{if } g_{ij} > 0 \end{cases}
$$
$$
\rho_{ij}^{-'} = \begin{cases} \rho_{ij}^{-}/\nu, & \text{if } g_{ij} > 0; \\ \rho_{ij}^{-} \cdot \nu, & \text{if } g_{ij} < 0 \end{cases}
\tag{3}
$$

With the adjustment by (3), there is an obvious gap between $\rho_{ij}^{+'}$ and $\rho_{ij}^{-'}$ for each element, in order to make modification tend to the expected direction. However, it is not suitable for non-feedback scheme because it lacks of mechanism to control the amount of adjusted cost. A large amount of adjusted cost has a negative effect.

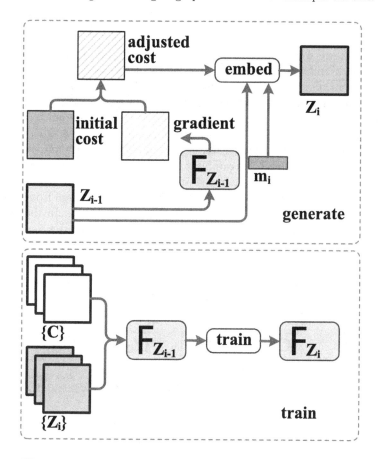

Fig. 2. Training the model $F_{\mathbf{Z}_i}$ via 'training-generating' process.

The adjustment scheme proposed in [4] takes the magnitude of quantization residual into account. Inspired by [4], the proposed cost adjustment scheme further exploits gradient. The magnitude of gradient reflects the impact on the inference result. Considering the magnitude, the expression of new cost adjustment scheme is (4).

$$\begin{cases} \rho_{ij}^{+'} = (1 + g_{ij}) \cdot \rho_{ij}^{+}; \\ \rho_{ij}^{-'} = (1 - g_{ij}) \cdot \rho_{ij}^{-} \end{cases} \quad (4)$$

In (4), g_{ij} is normalized to $[-1, 1]$ by $g_{ij} = g_{ij}/max(\|\mathbf{G}\|)$. The gradient signal g_{ij} controls the intensity of adjustment. The performance of 2 adjustment schemes are tested in experiment and (4) performs better.

3.4   Process of SAE

In SAE, there are 4 partial embedding steps to generate an adversarial stego sample. In each embedding step there is a particular adversary model to calculate the gradient. For example, the adversarial gradient used for generating $\mathbf{Z}_i$ is calculated from model $F_{\mathbf{Z}_{i-1}}$. Figure 3 is the flow chart of SAE, which displays the data and model in each step. The unnecessary elements are omitted in Fig. 3, i.e., message subsequence. The upper part is the training process. The lower part is the generating process. In the figure, arrow colored in red is embedding operation, arrow colored in green is training operation. In the training process, the conventional stego set $\{\mathbf{S}_1\}$ is generated at first. $\mathbf{S}_1$ is the stego image that carries the first 1/4 message segment. Then the model $F_{\mathbf{S}_1}$ is trained from $\{\mathbf{C}\}$ and $\{\mathbf{S}_1\}$. The following training process is introduced in Sect. 3.2. In the generating process, partial embedding is performed with the adversaries generated in training process.

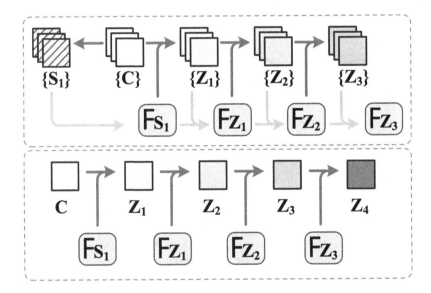

Fig. 3. The flow chart of SAE.

Besides the SAE displayed in Fig. 3, there is another simplified variant, which is illustrated in Fig. 4. Considering the partial embedding in [19], the first part of message is embedded in the cover with conventional scheme. In simplified SAE, the partial embedding of first 1/4 meassge segment is conventional embedding. The simplified variant is named as SAE-QC(SAE Quarter Conventional). SAE-QC has less storage requirement (3 models vs. 4 models) and less running time (times of gradient calculation).

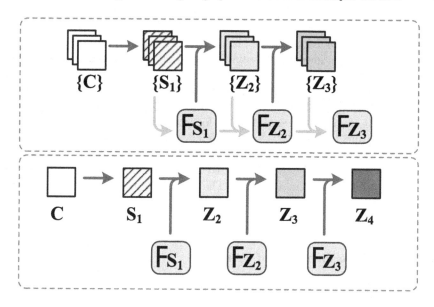

Fig. 4. The flow chart of SAE-QC

4 Experiments

4.1 Setup

The experiments, we choose UERD [10] as the distortion function to calculate the initial cost, because it has less complexity comparing with J-UNIWARD [11] and similar security performance.

The adversary model is an important component in SAE. In ALASKA2 steganalysis competition [5], EfficientNet [18] is widely used and proven well-performed. In this paper, we employ a modified version of EfficientNet-B2, whose input is single channel image and output is 2 category. As the input of the original version of EfficientNet is RGB 3-channel image, we triple the single channel input to adapt the EfficientNet structure. The model is fine-tuned from ImageNet pre-trained model. Note that input of the neural network is spatial image acquired from IDCT transformation without rounding operation.

The gradient map calculated from CNN model reflects the gradients in spatial domain, which cannot be used for cost adjustment in DCT domain directly. The spatial gradient is needed to be transformed to DCT gradient by (5).

$$\mathbf{G}_{dct8} = \mathbf{A}\mathbf{G}_{spa8}\mathbf{A}^T \tag{5}$$

In (5), $\mathbf{G}_{spa8}$ is 8×8 gradient block in spatial domain, while $\mathbf{G}_{dct8}$ is gradient block in DCT domain. $\mathbf{A}$ and $\mathbf{A}^T$ are the 2-D blockwise DCT transformation matrixes. (5) is the typical DCT transformation. After the transformation (5), $\mathbf{G}_{dct8}$ is divided by the block quantization matrix $\mathbf{Q}_8$ (6).

$$\mathbf{G}_{dct8} = \mathbf{G}_{dct8}./\mathbf{Q}_8 \tag{6}$$

For the image database, we choose official ALASKA2 JPEG base. The number of images is 80005. The image size is 256. The quality factor is 75. 40000 images are randomly selected as training set of adversary models. The rest of images are used for evaluation, 20000 are for training steganalyzers, 20005 are for testing. Stego image is generated via embedding simulator. The total payload rate is set to 0.4 bpnzac.

Three kinds of steganalyzer are chosen for evaluation, which are EfficientNet-B2, SRNet [3] and DCTR [12] + ensemble classifier [13]. EfficientNet-B2 is for evaluation of targeted steganalysis, SRNet and DCTR + ensemble classifier are for non-targeted steganalysis. CNN-based steganalyzers (EfficientNet-B2 and SRNet) are implemented in PyTorch. The optimizer is AdamW and the initial learning rate is set to 0.001. Batch size is set to 64 and epoch number is set to 50.

4.2   Security Performance of SAE

In this section, the security performance of SAE is evaluated by steganalyzer. The result is shown in Table 1. Comparing with conventional UERD, UERD-SAE decreases the detection accuarcy for 8.5%, 7.32%, 4.56% on EfficientNet, SRNet and DCTR respectively. There are more improvements in CNN-based steganlyzers than feature-based steganalyzers.

Table 1. Detection Accuracy evaluated on different steganalyzers. SAE is compared with conventional UERD

	EfficientNet	SRNet	DCTR
UERD	85.21%	85.96%	74.60%
UERD-SAE	76.71%	78.64%	69.95%

4.3   Security Performance of SAE-QC

The performance of SAE-QC is evaluated in this section. Table 2 is the result. For feature-based steganalyzer (DCTR), UERD-SAE and UERD-SAE-QC is nearly the same, while there are some differences for EfficientNet and SRNet. Generally, UERD-SAE-QC has similar performance with UERD-SAE, and less complexity.

Table 2. Detection Accuracy evaluated on different steganalyzers. UERD-SAE-QC is compared with UERD-SAE

	EfficientNet	SRNet	DCTR
UERD-SAE	76.71%	78.64%	69.95%
UERD-SAE-QC	77.46%	77.51%	69.99%

4.4   Comparison of Cost Adjustment Schemes

In this section we compare 2 cost adjustment schemes discussed in Sect. 3.3. The first part is security evaluation. Table 3 is the result. As the cost adjustment scheme (3) utilizes constant as factor to adjust cost, it is noted as UERD-SAE-CA (Constant Adjustment). In comparison, scheme (4) is UERD-SAE. In scheme (3), the constant for adjustment ν is set to 2. From Table 3, it can be seen that UERD-SAE-CA performs even worse than conventional UERD. Without feedback mechanism, scheme (3) cannot improve the security performance.

Table 3. Detection Accuracy evaluated on different steganalyzers. UERD-SAE-CA, UERD-SAE and conventional UERD are in the evaluation

	EfficientNet	SRNet	DCTR
UERD	85.21%	85.96%	74.60%
UERD-SAE-CA	90.14%	90.49 %	76.61%
UERD-SAE	76.71%	78.64%	69.95%

We randomly select 100 images from testing set to calculate the relative modification rate of 2 cost adjustment schemes. The relative modification rate is calculated by (7).

$$relative\ modification\ rate = \frac{number\ of\ modifications}{number\ of\ nzac} \tag{7}$$

The relative modification rate of UERD-SAE is 0.0916, UERD-SAE-CA is 0.0955. Scheme (3) introduces more modifications. Figure 5 is the visualization of the embedding. We pick out '00042.jpg' from the testing set. Image in the left is the cover. Image in the middle is the changed map of UERD-SAE. Image in the right is the changed map of UERD-SAE-CA. The regions in yellow boxes are in smooth area of the image. The white points indicates the positions of modification. Note that their random seeds for embedding are equal. It can be seen in the boxes that there are more modifications in UERD-SAE-CA than in UERD-SAE. In UERD-SAE-CA, modifications "leak" into smooth area. The modification in the smooth area has negative impact on the security.

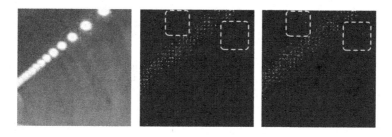

Fig. 5. The visualization of modifications in UERD-SAE and UERD-SAE-CA.

5 Conclusion

In this paper we propose a novel adversarial embedding scheme for JPEG steganography, which is named as Segmented Adversarial Embedding (SAE). The main contributions of our work including:

(1) Partial embedding is introduced into proposed scheme. In SAE, the message is divided into subsequence and embedded into cover in several separated processes. The adversarial embedding is performed in each partial embedding process. In each process, there is a specific adversary model for calculating the gradient to adjust the embedding cost.
(2) A novel cost adjustment scheme is applied in SAE. The scheme exploits both sign and magnitude of gradient signal to adjust embedding cost, rather than only considering the sign.
(3) A simplified version of SAE is developed. The simplified SAE performs the first segment embedding with conventional schemes rather than adversarial embedding. Comparing with the original SAE, the simplified SAE has less running complexity.

The future work will mainly focus on 2 aspects. First is to explore the lightweight scheme to adjust the amount of cost being adjusted, rather than performing embedding multiple times. Second is to explore the application of knowledge distilling in adversarial embedding.

References

1. Bernard, S., Bas, P., Klein, J., Pevny, T.: Explicit optimization of min max steganographic game. IEEE Trans. Inf. Forensics Secur. **16**, 812–823 (2021)
2. Bernard, S., Pevnỳ, T., Bas, P., Klein, J.: Exploiting adversarial embeddings for better steganography. In: Proceedings of the ACM Workshop on Information Hiding and Multimedia Security, pp. 216–221. ACM (2019)
3. Boroumand, M., Chen, M., Fridrich, J.: Deep residual network for steganalysis of digital images. IEEE Trans. Inf. Forensics Secur. **14**(5), 1181–1193 (2019)
4. Butora, J., Fridrich, J.: Steganography and its detection in JPEG images obtained with the "TRUNC" quantizer. In: ICASSP 2020–2020 IEEE International Conference on Acoustics, Speech and Signal Processing (ICASSP), pp. 2762–2766 (2020)
5. Cogranne, R., Giboulot, Q., Bas, P.: Alaska#2 steganalysis challenge. http://alaska.utt.fr
6. Denemark, T., Fridrich, J.: Improving steganographic security by synchronizing the selection channel. In: Proceedings of the 3rd ACM Workshop on Information Hiding and Multimedia Security, IH&MMSec 2015, pp. 5–14. ACM, New York (2015)
7. Filler, T., Judas, J., Fridrich, J.: Minimizing additive distortion in steganography using syndrome-trellis codes. IEEE Trans. Inf. Forensics Secur. **6**(3), 920–935 (2011)
8. Fridrich, J., Filler, T.: Practical methods for minimizing embedding impact in steganography. In: Delp III, E.J., Wong, P.W. (eds.) Security, Steganography, and Watermarking of Multimedia Contents IX, vol. 6505, pp. 13–27. International Society for Optics and Photonics, SPIE (2007)

9. Goodfellow, I., Shlens, J., Szegedy, C.: Explaining and harnessing adversarial examples. arXiv preprint arXiv:1412.6572 (2014)
10. Guo, L., Ni, J., Su, W., Tang, C., Shi, Y.: Using statistical image model for JPEG steganography: Uniform embedding revisited. IEEE Trans. Inf. Forensics Secur. **10**(12), 2669–2680 (2015)
11. Holub, V., Fridrich, J.: Digital image steganography using universal distortion. In: Proceedings of the First ACM Workshop on Information Hiding and Multimedia Security, pp. 59–68. ACM (2013)
12. Holub, V., Fridrich, J.: Low-complexity features for JPEG steganalysis using undecimated DCT. IEEE Trans. Inf. Forensics Secur. **10**(2), 219–228 (2015)
13. Kodovsky, J., Fridrich, J., Holub, V.: Ensemble classifiers for steganalysis of digital media. IEEE Trans. Inf. Forensics Secur. **7**(2), 432–444 (2012)
14. Kouider, S., Chaumont, M., Puech, W.: Adaptive steganography by oracle (ASO). In: 2013 IEEE International Conference on Multimedia and Expo (ICME), pp. 1–6, July 2013
15. Li, B., Wang, M., Li, X., Tan, S., Huang, J.: A strategy of clustering modification directions in spatial image steganography. IEEE Trans. Inf. Forensics Secur. **10**(9), 1905–1917 (2015)
16. Ma, S., Zhao, X., Liu, Y.: Adaptive spatial steganography based on adversarial examples. Multimed. Tools Appl. **78**(22), 32503–32522 (2019). https://doi.org/10.1007/s11042-019-07994-3
17. Mo, H., Song, T., Chen, B., Luo, W., Huang, J.: Enhancing jpeg steganography using iterative adversarial examples. In: 2019 IEEE International Workshop on Information Forensics and Security (WIFS), pp. 1–6 (2019)
18. Tan, M., Le, Q.V.: EfficientNet: rethinking model scaling for convolutional neural networks. arXiv preprint arXiv:1905.11946 (2019)
19. Tang, W., Li, B., Tan, S., Barni, M., Huang, J.: Cnn-based adversarial embedding for image steganography. IEEE Trans. Inf. Forensics Secur. **14**(8), 2074–2087 (2019)
20. Zhang, W., Zhang, Z., Zhang, L., Li, H., Yu, N.: Decomposing joint distortion for adaptive steganography. IEEE Trans. Circ. Syst. Video Technol. **27**(10), 2274–2280 (2017)
21. Zhang, Y., Zhang, W., Chen, K., Liu, J., Liu, Y., Yu, N.: Adversarial examples against deep neural network based steganalysis. In: Proceedings of the 6th ACM Workshop on Information Hiding and Multimedia Security, pp. 67–72. ACM (2018)

High-Performance Linguistic Steganalysis, Capacity Estimation and Steganographic Positioning

Jiajun Zou[1,2], Zhongliang Yang[1,2(✉)], Siyu Zhang[1,2], Sadaqat ur Rehman[3], and Yongfeng Huang[1,2]

[1] Beijing National Research Center for Information Science and Technology, Beijing, China
[2] Department of Electronic Engineering, Tsinghua University, Beijing 100084, China
{zoujj18,zhangsiy19}@mails.tsinghua.edu.cn, yangzl15@tsinghua.org.cn, yfhuang@tsinghua.edu.cn
[3] Department of Computer Science, Namal Institute, Mianwali 42250, Pakistan
sadaqat.rehman@namal.edu.pk

Abstract. With the rapid development of natural language processing technology, various linguistic steganographic methods have been proposed increasingly, which may bring great challenges in the governance of cyberspace security. The previous linguistic steganalysis methods based on neural networks with word embedding layer could only extract the context-independent word-level features, which are insufficient for capturing the complex semantic dependencies in sentences, thus may limit the performance of text steganalysis. In this paper, we propose a novel linguistic steganalysis model. We first employ the BERT or Glove component to extract the contextualized association relationships of words in the sentences. Then we put these extracted features into BiLSTM to further get context information. We use the attention mechanism to find out local parts that may be discordant in text. Finally, based on these extracted features, we use the softmax classifier to decide if the input sentence is cover or stego. Experimental results show that the proposed model can achieve currently the best performance of text steganalysis and hidden capacity estimation. Further experiments found that proposed model can even locate where the secret information may be embedded in the text to a certain extent. To the best of our knowledge, we made the first attempt to achieve text steganography positioning in the field of text steganalysis (Code and datasets are available at https://github.com/YangzlTHU/Linguistic-Steganography-and-Steganalysis).

Keywords: Text steganalysis · Capacity estimation · Hidden positioning

Supported by the National Key Research and Development Program of China under Grant No. 2018YFB0804103 and the National Natural Science Foundation of China (No. U1936208, No. 61862002 and No. U1936216).

X. Zhao et al. (Eds.): IWDW 2020, LNCS 12617, pp. 80–93, 2021.
https://doi.org/10.1007/978-3-030-69449-4_7

1   Introduction

Nowadays, people are used to share and transmit information conveniently on the Internet with the development of networks. In the meantime, the challenges of protecting information security and privacy have been paid attention by many researchers [24]. According to Claude E. Shannon [17], encryption system, privacy system, and concealment system are the three main information secrecy systems in cyberspace. The biggest difference between the concealment system and the other two systems is the concealment system can hide the existence of important information in the form of embedding them into common carriers, ensuring its security. The history of people using steganography to transmit secret information on various occasions can even be traced back to the 14th and 15th centuries. Through steganography, secret messages are embedded into cover carriers such as image [22,31], audio [28], text [6,11,27,32,34–36] and so on [3,16]. Text is one of the most important information carriers in human lives [12]. The extensive text interaction scenarios and massive text carriers on the Internet have attracted more and more researchers in recent years to study text steganography and use network texts to implement covert communication. However, text steganography can also be used by hackers, terrorists, and other law breakers for illegitimate purposes, which threats cybersecurity seriously. Therefore, it is crucial to develop an effective and powerful text steganalysis method.

A concealment system can be illustrated by Simmons' "Prisoners' Problem" [18]. The core goal of steganography is to reduce the difference of statistical distribution between carriers before and after steganography as much as possible. However, steganalysis aims to recognize the difference to the greatest extent possible.

Most of text steganography can be divided into two types, modification-based [4] and generation-based [11,28,32]. Modification-based methods usually embed secret information by modifying the cover texts, such as synonym substitution, etc. [5,15]. Generation-based text information hiding method can automatically generate steganographic texts according to the secret information [6,27,32,34,36].

Steganalysis of text always follows the same framework: extracting statistical features from texts and then send them to a specific classifier to determine whether they are normal texts or steganographic texts. Traditional text steganalysis methods usually construct simple text statistical features manually [20], and then analyze the difference between normal texts and steganographic texts, which is slow and not efficient enough. In recent years, with the development of artificial neural networks (ANN) and natural language processing (NLP) technology, more and more automatic steganographic texts generation models based on neural networks have emerged, which have been able to generate steganographic sentences with high quality [6,8,27,32,34,36]. So proposing stronger steganalysis approaches to respond to the rapid developments in linguistic steganography is very important. Due to the development of neural networks, many text steganalysis methods have also introduced neural network models to achieve better detection results [1,23,25,29]. All these models roughly contain a word embedding module, a hidden module and a final classification module. Although these models can achieve a good steganalysis performance [1,25], it is seems that they reach the ceiling, and the improvement in detection performance is slow, such as [1,25]. One reason is that steganography could influence the correlations between words thus changing the semantic features in texts [30], the currently used word embedding layer, trained by the model self, only provides the context-independent word-level features, which is not enough to obtain the complex semantic dependencies in sentences.

In this paper, we first adopted BERT (Bidirectional Encoder Representation from Transformers) [7], one of the most popular pre-trained language model armed with Transformer [21], as an embedding layer, which can learn richer features in the sentences, and Glove (Global Vectors for Word Representation) [14] was also used as the embedding layer as a comparison. In order to fuse contextual information, we used BiLSTM (Bidirectional Long Short-Term Memory) after the output of embedding layer. Furthermore, the attention mechanism was also introduced in the proposed model for focusing on the suspicious parts in sentences. Experimental results show that the proposed model greatly exceeds all the previous models in text steganalysis and hidden information capacity estimation tasks, which show the state-of-the-art performance. Further experiments found that proposed model can even locate where the secret information may be embedded in the texts to a certain extent. To the best of our knowledge, we made the first attempt to achieve text steganography positioning in the field of text steganalysis.

In the remainder of this paper, Sect. 2 introduces related works on text steganalysis. We explain the proposed model in detail in Sect. 3. Section 4 reveals on the experiments results and gives a comparison between different models. Finally, conclusions are drawn in Sect. 5.

2   Related Works

In this section, we will first briefly describe the principle of text steganalysis, and then introduce some existing text steganalysis works proposed in recent years.

In the field of natural language processing, a sentence can be considered as a sequence signal. In the purpose of getting a semantically complete word sequence, the statistical language model [2] is the most common approach, which learns the conditional probability of each word in normal sentences, it can be expressed in the following formula:

$$\begin{aligned} p(X) &= p(x_1, x_2, \ldots, x_n) \\ &= p(x_1)p(x_2|x_1)\ldots p(x_n|x_{n-1}), \end{aligned} \tag{1}$$

where X denotes the whole sentence, which has n words, and x_i is the i-th word in it.

In recent years, many text steganography methods including modification or generation forms, first analyze the statistical feature distribution in natural texts by using neural networks, and then reconstruct steganographic texts to reduce the statistical distribution differences with the natural texts. Thus, the key task of text steganalysis is to recognize these changes of the statistical features such as the conditional probability.

There have been many methods in text steganalysis. For instance, in 2006, Support Vector Machine (SVM) was used to distinguish the stego texts modified by a lexical steganography algorithm [20]. In 2010, Yang and Cao [26] proposed a novel linguistic steganalysis method through meta features and immune clone mechanism. These approaches always use handcrafted features which are labor intensive.

With the development of deep learning, neural networks have shown huge advantages in many tasks. Some researchers have paid attention to using neural networks to deal with text steganalysis. In 2019, Yang et al. [29] proposed a fast and efficient text steganalysis method which mapped words into a semantic space and used a hidden layer to extract the correlations between these words by using Recurrent Neural Network (RNN). Wen et al. [23] used Convolutional Neural Networks (CNN) to do text steganalysis, and a word embedding layer was used to extract the semantic and syntax features of words. In order to capture both local and long-distance contextual information in steganography texts, Bao et al. [1] combined CNN and Long Short-Term Memory (LSTM) [9] recurrent neural networks, and got the state-of-art results in the text steganalysis task.

As a matter of fact, in the generated steganographic texts, the word correlation features may be distorted after being embedded with secret information [30], resulting in the changes of the semantic relations in natural texts, which helps us to achieve text steganalysis in the way of extracting and analyzing the word correlations and semantic features related to words of the steganographic texts in natural sentences. As mentioned above, there exist limitations in previous methods actually, that is they just use one word embedding layer to extract text features, then fuse and analyze these features. However, a word embedding layer can not fully represent more features, such as complex semantic relations, dependencies between sentences and so on. These place restrictions on performance of text steganalysis methods. Thus, for the purpose of extracting more features not only at the level of words, but also at the semantic level, we take the BERT, a pre-trained model, as the embedding layer, because of its powerful feature representation ability, and Glove acts as a comparison. Furthermore, we use the attention mechanism to locate the secret informations and heatmaps are drawn.

3   The Proposed Model

The text steganalysis task can be formulated as a text classification problem. As shown above, steganographic texts embedded secret information will influence the correlations between words in sentences, further leading to change semantic features. So if we can extract more informations other than word level, we can reach better results in text steganalysis task. Taking into account this fact, we proposed a novel model, the overall architecture of which is depicted in Fig. 1.

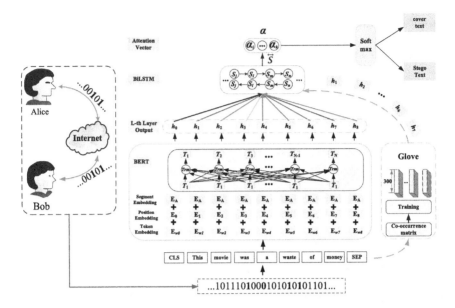

Fig. 1. The overall framework of the proposed text steganalysis method. We use BERT to extract more features (Glove for comparison), and we use BiLSTM further to fuse context informations, the attention mechanism focuses on important parts.

3.1   Extracting Semantic Features

The BERT embedding layer takes the sentence as input and calculates the token-level representations using the information from the whole sentence. Given the input token sequence $X = \{x_1, \ldots, x_n\}$ of length n, we firstly employ BERT component with L transformer layers to calculate the contextualized representations $H^L = \{h_1^L, \ldots, h_n^L\} \in \mathbb{R}^{n \times dim_h}$ for the input tokens, dim_h denotes the dimension of the representation vectors. The input feature is $H^0 = \{e_1, \ldots, e_n\}$, where the token embedding, position embedding and segment embedding corresponding to the input token x_t combine $e_t (\mathrm{t} \in [1, n])$. And the L transformer layers will refine the token-level features layer by layer. The l-th layer representation is calculated below:

$$H^l = Transformer_l(H^{l-1}) \tag{2}$$

where $H^l = \{h_1{}^l, \ldots, h_n{}^l\}$. We use H^L to do the downstream task.

In addition, we use Glove as a control, which efficiently leverages statistical information by training only on the nonzero elements in a word-word co-occurrence matrix, rather than on the entire sparse matrix or on individual context windows in a large corpus [14].

3.2   Feature Fusion

LSTM [9] is a variant of RNN which is very suitable for modeling sequential signals such as text, it is good at learning long-term dependencies and avoiding gradient vanishing and expansion problems. The input of LSTM is the output of BERT, which is described as $h_k{}^L$, where k is the k-th token in the sentence, if c_{k-1} denotes the previous cell state and previous hidden state is s_{k-1}, then the current cell state c_k and current hidden state s_k are updated as follows:

$$
\begin{aligned}
\mathbf{i}^k &= \sigma\left(W_\mathbf{i}^w \cdot w^k + W_\mathbf{i}^h \cdot h^{k-1} + b_\mathbf{i}\right) \\
\mathbf{f}^k &= \sigma\left(W_\mathbf{f}^w \cdot w^k + W_\mathbf{f}^h \cdot h^{k-1} + b_\mathbf{f}\right) \\
\mathbf{o}^k &= \sigma\left(W_\mathbf{o}^w \cdot w^k + W_\mathbf{o}^h \cdot h^{k-1} + b_\mathbf{o}\right) \\
\hat{c}^k &= \tanh\left(W_c^w \cdot w^k + W_c^h \cdot h^{k-1} + b_\mathbf{c}\right) \\
c^k &= \mathbf{f}^k \odot c^{k-1} + \mathbf{i}^k \odot \hat{c}^k \\
s^k &= \mathbf{o}^k \odot \tanh\left(c^k\right)
\end{aligned}
\tag{3}
$$

where $\mathbf{i}$, $\mathbf{f}$ and $\mathbf{o}$ are input gate, forget gate and output gate respectively, σ is the sigmoid function and $\odot$ stands for element-wise multiplication, and the last hidden state vector s_k is the representation of the sentence. In order to get bidirectional information, we concatenate the forward hidden state vector $\overrightarrow{s_k}$ and the backward hidden state vector $\overleftarrow{s_k}$ as the last output of BiLSTM at step k,

$$
\overleftrightarrow{S_k} = [\overrightarrow{s_k}, \overleftarrow{s_k}]
\tag{4}
$$

Thus, $\overleftrightarrow{S} = [\overleftrightarrow{S_1}, \ldots, \overleftrightarrow{S_n}] \in \mathbb{R}^{2d \times |n|}$, d is the hidden size of LSTM, we can abbreviate the computation of the BiLSTM layer with $\overleftrightarrow{S} = BiLSTM(H^L)$.

3.3   Attention Layer

Attention mechanism [21] can capture the important information in sentences, so we set an attention layer after the BiLSTM layer. As shown above, the output of BiLSTM $\overleftrightarrow{S}$ is the input of the attention layer, $s_k \in \overleftrightarrow{S}$, then we calculate the self attention vector α_i as follows:

$$
\begin{aligned}
Q &= W_Q X \in \mathbf{S}^{d \times n} \\
K &= W_K X \in \mathbf{S}^{d \times n} \\
V &= W_V X \in \mathbf{S}^{d \times n}
\end{aligned}
\tag{5}
$$

$$
\begin{aligned}
e_{ij} &= v_\alpha^T tanh(W_a s_{i-1} + U_a h_j) \\
\alpha_{ij} &= \frac{\exp\left(e_{ij}\right)}{\sum_{k=1}^{T_x} \exp\left(e_{ik}\right)}
\end{aligned}
\tag{6}
$$

where we use a Multilayer Perceptron (MLP) [13] as the score function to calcuLate the attention score. And Q is the query vector sequence, K is key vector sequence, and V is value vector sequence. W_Q, W_K, W_V, W_a and U_a are the parameters of the attention layer. Finally we can get the attention matrix.

3.4   Classification Layer

As text steganalysis is a sentence classification task, we calculate the average attention score of the whole sentence rather than fed the token's attention score to a fully-connected layer. We designed a softmax normalization layer to yield a probability distribution $\mathbf{p} \in \mathbb{R}^{d_p}$ over the decision space:

$$p = softmax(W_p \alpha_{ij} + b_p) \tag{7}$$

where d_p is the same as the number of labels, W_p and b_p are the learned weights and bias respectively.

3.5   Training

The model is trained by minimizing the cross entropy and the loss function is expressed as following formula:

$$loss = -\Sigma_i \Sigma_j y_i^j log \hat{y}_i^j + \lambda ||\theta||^2 \tag{8}$$

where λ is the coefficient for L2-regularization, θ denotes the parameters that need to be regularized, y denotes the true label, $\hat{y}$ is the predicted label.

4   Experiments

4.1   Datasets and Experimental Settings

We use Twitter and News text media from the T-Steg dataset collected by Yang et al. [33]. The natural steganographic texts were generated by Fang et al. [8] with different embedding rates by altering the number of bits hidden per word (bpw, bits per word). The details of Twitter and News datasets are described in Table 1. In the purpose of training and testing the proposed model, we firstly mix up the Cover-Texts and Stego-texts with different embedding rates respectively, and then randomly select 60% of the sample for training, 20% of the samples for validation, and the rest 20% of samples for testing. One goal of the proposed model is to classify the Cover-Texts and Stego-texts.

For our experiments, we use the pre-trained "bert-base-uncased" model, where the number of transformer layers is $L = 12$, the hidden size dim_h is 768. The max length of sentences is set to be 128, we set the batch size to 32, the learning rate is 5e−5. Apart from this, we apply dropout [19] and L2-regularization, the dropout rate is 0.1 and the coefficient rate λ of L2 is 1e−5. We train the model for 20 epochs to get the best results. We used AdamW [10] as the optimization method.

On the other hand, as comparison, we exchange BERT in the proposed model with Glove[1] to examine the proposed model's performance, because Glove is a more powerful model than Word2Vec, thus we can further verify if our model can capture more features. For this, we set BiLSTM hidden size to be 300, batch size to be 128, and adjust learning rate to be 1e−3.

---

[1] Pre-trained word embedding of GloVe can be downloaded from http://nlp.stanford.edu/projects/glove/.

Table 1. The details of the training datasets [33]

Dataset	Average length	Sentence number	Total words
Twitter	9.68	2,639,290	46,341
News	22.24	1,962,040	42,745

4.2   Evaluation Metrics

In the experiment, we use several evaluation indicators which are commonly used in text steganalysis task to evaluate the performance of the proposed model. These metrics are Accuracy (Acc), Precision (P) and Recall (R), which formulas are described below:

$$Accuracy = \frac{TP + TN}{TP + FP + TN + FN}$$

$$Precision = \frac{TP}{TP + FP} \qquad (9)$$

$$Recall = \frac{TP}{TP + TN}$$

in the above formulas, TP (True Positive) represents the number of positive samples which are predicted to be positive, FP (False Positive) is the number of negative samples predicted to be positive, and the number of positive samples predicted to be negative is FN (False Negative), TN (True Negative) illustrates the number of negative samples predicted to be negative.

4.3   Models for Comparison

In order to comprehensively evaluate the proposed model, we compare it with a range of baselines and state-of-the-art models, in addition, we also exchange BERT with Glove [14] as the embedding layer to further verify the performance of the proposed model:

1. **A Fast and Efcient Text Steganalysis Method** [29]: first analyzed the correlations between words in steganographic texts, then mapped each word to a semantic space and used a hidden layer to extract the correlations.
2. **CNN based text steganalysis** [23]: adopt CNN to capture complex dependencies and learn feature representations automatically from the texts.
3. **Text Steganalysis with Attentional LSTM-CNN** [1]: combined CNN and LSTM to capture both local and long-distance contextual information in steganographic texts, it also used an attention mechanism.
4. **Linguistic Steganalysis via Densely Connected LSTM with Feature Pyramid** [25]: proposed densely connected LSTM with feature pyramids which can incorporate more low level features to recognize steganographic texts.

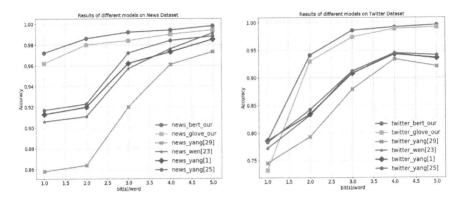

Fig. 2. The changes of text steganalysis performance of the different models on News and Twitter Datasets with the increase of embedding rate

4.4 Experiment Results

Table 2 shows the performance comparison of the proposed model with other models in the text steganalysis task under different embedding rate (*bpw*). Table 3 shows the results of secret information estimation task. We first mixed the texts of various embedding rates, then we used the proposed model to predict the embedding rate of the information hiding in each text, and record the prediction accuracy as shown in Table 3. According to the results, we can draw several conclusions as following.

Firstly, the proposed model has achieved the best detection results on various metrics, no matter on News dataset or Twitter dataset. And the evaluation indicators have been greatly improved, especially at low *bpw* on News dataset and higher *bpw* on Twitter dataset as shown in Fig. 2. The reason could be that News texts are more regular and longer than Twitter texts, which makes the proposed model capture more features than other models even though at low *bpw*. On the other hand, Twitter texts are anomalistic and shorter than News texts, resulting in the difficulty in detection at low *bpw*, but as the *bpw* in Twitter texts increase, the performance of the proposed model has been significantly improved than other models. Figure 2 clearly shows these results.

Secondly, it is obvious that the detection performance of all models have improved with the increasing of the embedding rate. This is easy to understand, because once more information is embedded in texts, coherence of text semantics and statistical distribution features will be changed more.

Thirdly, from Table 3, in the task of hidden information capacity estimation, we can see the accuracies of the proposed model have achieved a higher level especially BERT, reaching up to **93.83%** and **85.74%** in News and Twitter datasets respectively, which are more excellent than other models [30]. This means that the proposed model can estimate the capacity of hidden information inside the natural texts more excellent.

Table 2. Results of different steganalysis methods on different datasets.

Methods		News [33]					Twitter [33]				
Yang et al. 2019 [29]	bpw	1	2	3	4	5	1	2	3	4	5
	Acc	0.858	0.864	0.920	0.961	0.973	0.745	0.793	0.879	0.934	0.921
	P	0.858	0.915	0.922	0.979	0.988	0.811	0.914	0.939	0.988	0.960
	R	0.858	0.803	0.918	0.942	0.958	0.621	0.647	0.812	0.879	0.879
Wen et al. 2019 [23]	bpw	1	2	3	4	5	1	2	3	4	5
	Acc	0.906	0.911	0.957	0.976	0.991	0.772	0.832	0.912	0.944	0.935
	P	0.906	0.916	0.955	0.983	0.996	0.854	0.877	0.933	0.968	0.921
	R	0.905	0.904	0.959	0.968	0.985	0.640	0.772	0.888	0.917	0.950
Yang et al. 2019 [1]	bpw	1	2	3	4	5	1	2	3	4	5
	Acc	0.913	0.920	0.962	0.973	0.985	**0.786**	0.834	0.908	0.943	0.936
	P	0.930	0.923	0.966	0.981	0.983	0.873	0.883	0.950	0.986	0.958
	R	0.894	0.916	0.958	0.966	0.987	0.657	0.770	0.861	0.899	0.911
Yang et al. 2020 [25]	bpw	1	2	3	4	5	1	2	3	4	5
	Acc	0.917	0.923	0.972	0.984	0.988	0.783	0.842	0.912	0.945	0.941
	P	0.922	0.933	0.974	0.989	0.989	0.817	0.884	0.907	0.964	0.958
	R	0.910	0.913	0.969	0.979	0.987	0.714	0.786	0.919	0.925	0.923
Glove -BiLSTM-Att	bpw	1	2	3	4	5	1	2	3	4	5
	Acc	0.962	0.980	0.984	0.990	0.994	0.732	0.930	0.974	0.989	0.992
	P	0.972	0.976	0.983	0.993	0.995	0.835	0.944	0.978	0.994	0.995
	R	0.961	0.988	0.988	0.990	0.994	0.743	0.947	0.982	0.989	0.992
BERT -BiLSTM-Att	bpw	1	2	3	4	5	1	2	3	4	5
	Acc	**0.972**	**0.986**	**0.992**	**0.994**	**0.998**	**0.786**	**0.941**	**0.986**	**0.992**	**0.996**
	P	**0.974**	**0.992**	**0.996**	**0.999**	**0.998**	**0.914**	**0.966**	**0.991**	**0.997**	**1.000**
	R	**0.977**	**0.983**	**0.991**	**0.990**	**0.998**	**0.744**	**0.945**	**0.989**	**0.992**	**0.995**

In addition to the excellent performance in the above-mentioned text steganalysis and hidden information capacity estimation tasks, we also tried to explore the unique performance of the proposed model. For example, we found that it can locate which words are more likely to be embedded in the text to a certain extent. We first used the method proposed by Fang et al. [8] to generate some new sentences. In the process of generating these sentences, we randomly selected a small number of words in the text to embed information with random capacity (*bpw* was randomly set to 1–5), and the remaining words did not embed information. Then we input these texts and normal texts into the proposed model for steganalysis. When the model completed the judgment of whether the input text was cover or stego, we extracted its attention weight vector, which reflected the degree of attention the model pays to each word in the input text. Finally, we visualized these results and shown in Fig. 3. The sentence on the left in Fig. 3 represents the input text, the red words are embedded with hidden information, and the black words have no embedded information. On the right is the visualization result of the model's attention weight. The darker the color, the larger the attention weight value, which means that for the text steganalysis character, the model believes that this local area is worthy of attention.

Table 3. The results of hidden information capacity estimation of the proposed model

Method		News [33]					Twitter [33]				
Glove-BiLSTM-Att	bpw	1	2	3	4	5	1	2	3	4	5
	Acc	**90.73**	87.58	89.95	91.72	94.47	67.84	80.63	85.57	87.96	95.40
	P	92.62	89.61	88.07	89.77	94.14	**75.26**	77.71	86.58	92.04	91.12
	F1	91.66	88.58	89.00	90.74	94.30	71.36	79.14	86.07	89.96	93.214
	Acc	91.43					81.30				
BERT -BiLSTM-Att	bpw	1	2	3	4	5	1	2	3	4	5
	Acc	89.04	**91.30**	**91.48**	**96.30**	**99.45**	**75.98**	**85.85**	**92.26**	**91.89**	**98.95**
	P	**94.52**	**91.93**	**93.51**	**93.91**	**95.13**	73.50	**86.79**	**89.81**	**95.21**	**92.74**
	F1	**91.70**	**91.62**	**92.48**	**95.09**	**97.24**	**74.72**	**86.32**	**91.02**	**93.52**	**95.75**
	Acc	**93.83**					**85.74**				

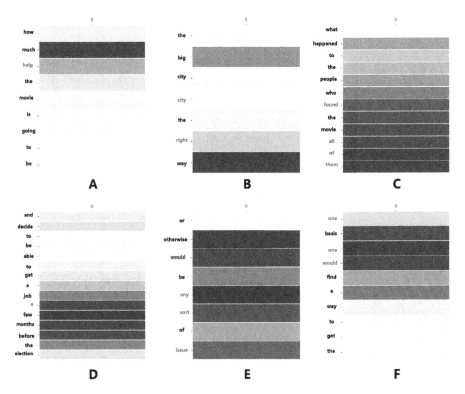

Fig. 3. The figure describes the locations of secret informations in the sentences. The real positions of cover words in sentences are marked red. The darker the color, the greater the attention score. We can see the proposed model located the secret informations effectively. (Color figure online)

By analyzing the results in Fig. 3, we can see some very interesting and valuable results. We found that the proposed model will give a relatively large attention weight to the local area where the secret information is embedded in the input text, which shows that the model can indeed locate the area where the secret information is embedded in the text to a certain extent. As far as we are concerned, this should be the first attempt to locate the location of embedded hidden information in the field of text steganalysis. Secondly, we found that the attention weight of the proposed model does not all focus on the words embedded in the secret information. In fact, it is like a "ripple" which gradually spreads to the surrounding area with the embedded information as the core. This further validates our previous analysis and our research motivation, that is, when we embed secret information in a word in the text, it will affect the relationship between the words around the entire word. The change of this association relationship can be used as an analysis feature to achieve high accurate text steganalysis after extraction.

5 Conclusion

With the developments of natural language processing technology, text steganalysis has faced many challenges, especially when neural networks become popular. Text steganography will change the correlations between words, thus we can do texts steganalysis. However the embedding layers using in many texts steganalysis models can not obtain complex semantic features in sentences which limits the performance of many methods. In this paper, we first use BERT or Glove as a embedding layer rather than traditional word embedding layers to capture more features, then we use a BiLSTM and the attention mechanism to obtain the important information in texts to do text steganalysis task. Experiments on different datasets are conducted by using the proposed model, the results show that the proposed scheme outperforms the state-of-the-art methods. Furthermore, we try for the first time to visualize attention weights in text steganalysis task, results show that our model can locate the secret information in steganographic texts.

References

1. Bao, Y., Yang, H., Yang, Z., Liu, S., Huang, Y.: Text steganalysis with attentional LSTM-CNN. arXiv preprint arXiv:1912.12871 (2019)
2. Bengio, Y., Ducharme, R., Vincent, P., Jauvin, C.: A neural probabilistic language model. J. Mach. Learn. Res. **3**(Feb), 1137–1155 (2003)
3. Boukis, A.C., Reiter, K., Frölich, M., Hofheinz, D., Meier, M.A.: Multicomponent reactions provide key molecules for secret communication. Nat. Commun. **9**(1), 1–10 (2018)
4. Chang, C.Y., Clark, S.: Practical linguistic steganography using contextual synonym substitution and vertex colour coding. In: Proceedings of the 2010 Conference on Empirical Methods in Natural Language Processing, pp. 1194–1203 (2010)
5. Chang, C.Y., Clark, S.: Practical linguistic steganography using contextual synonym substitution and a novel vertex coding method. Comput. Linguist. **40**(2), 403–448 (2014)

6. Dai, F.Z., Cai, Z.: Towards near-imperceptible steganographic text. arXiv preprint arXiv:1907.06679 (2019)
7. Devlin, J., Chang, M.W., Lee, K., Toutanova, K.: BERT: pre-training of deep bidirectional transformers for language understanding. arXiv preprint arXiv:1810.04805 (2018)
8. Fang, T., Jaggi, M., Argyraki, K.: Generating steganographic text with LSTMs. arXiv preprint arXiv:1705.10742 (2017)
9. Hochreiter, S., Schmidhuber, J.: Long short-term memory. Neural Comput. $9(8)$, 1735–1780 (1997)
10. Loshchilov, I., Hutter, F.: Fixing weight decay regularization in Adam (2018)
11. Luo, Y., Huang, Y., Li, F., Chang, C.: Text steganography based on Ci-poetry generation using Markov chain model. TIIS $10(9)$, 4568–4584 (2016)
12. Michel, J.B., et al.: Quantitative analysis of culture using millions of digitized books. Science $331(6014)$, 176–182 (2011)
13. Pal, S.K., Mitra, S.: Multilayer perceptron, fuzzy sets, classification (1992)
14. Pennington, J., Socher, R., Manning, C.D.: GloVe: global vectors for word representation. In: Proceedings of the 2014 Conference on Empirical Methods in Natural Language Processing (EMNLP), pp. 1532–1543 (2014)
15. Rizzo, S.G., Bertini, F., Montesi, D.: Content-preserving text watermarking through unicode homoglyph substitution. In: Proceedings of the 20th International Database Engineering & Applications Symposium, pp. 97–104 (2016)
16. Sarkar, T., Selvakumar, K., Motiei, L., Margulies, D.: Message in a molecule. Nat. Commun. $7(1)$, 1–9 (2016)
17. Shannon, C.E.: Communication theory of secrecy systems. Bell Syst. Tech. J. $28(4)$, 656–715 (1949)
18. Simmons, G.J.: The prisoners' problem and the subliminal channel. In: Chaum, D. (ed.) Advances in Cryptology, pp. 51–67. Springer, Boston (1984). https://doi.org/10.1007/978-1-4684-4730-9_5
19. Srivastava, N., Hinton, G., Krizhevsky, A., Sutskever, I., Salakhutdinov, R.: Dropout: a simple way to prevent neural networks from overfitting. j. Mach. Learn. Res. $15(1)$, 1929–1958 (2014)
20. Taskiran, C.M., Topkara, U., Topkara, M., Delp, E.J.: Attacks on lexical natural language steganography systems. In: Security, Steganography, and Watermarking of Multimedia Contents VIII, vol. 6072, p. 607209. International Society for Optics and Photonics (2006)
21. Vaswani, A., et al.: Attention is all you need. In: Advances in Neural Information Processing Systems, pp. 5998–6008 (2017)
22. Wang, Y., Zhang, W., Li, W., Yu, X., Yu, N.: Non-additive cost functions for color image steganography based on inter-channel correlations and differences. IEEE Trans. Inf. Forensics Secur. 15, 2081–2095 (2019)
23. Wen, J., Zhou, X., Zhong, P., Xue, Y.: Convolutional neural network based text steganalysis. IEEE Signal Process. Lett. $26(3)$, 460–464 (2019)
24. Wulf, W.A., Jones, A.K.: Reflections on cybersecurity. Science $326(5955)$, 943–944 (2009)
25. Yang, H., Bao, Y., Yang, Z., Liu, S., Huang, Y., Jiao, S.: Linguistic steganalysis via densely connected LSTM with feature pyramid. In: Proceedings of the 2020 ACM Workshop on Information Hiding and Multimedia Security, pp. 5–10 (2020)
26. Yang, H., Cao, X.: Linguistic steganalysis based on meta features and immune mechanism. Chin. J. Electron. $19(4)$, 661–666 (2010)

27. Yang, Z., Guo, X., Chen, Z., Huang, Y., Zhang, Y.: RNN-stega: linguistic steganography based on recurrent neural networks. IEEE Trans. Inf. Forensics Secur. **14**(5), 1280–1295 (2019). https://doi.org/10.1109/TIFS.2018.2871746
28. Yang, Z., Du, X., Tan, Y., Huang, Y., Zhang, Y.J.: AAG-stega: automatic audio generation-based steganography. arXiv preprint arXiv:1809.03463 (2018)
29. Yang, Z., Huang, Y., Zhang, Y.J.: A fast and efficient text steganalysis method. IEEE Signal Process. Lett. **26**(4), 627–631 (2019)
30. Yang, Z., Huang, Y., Zhang, Y.J.: TS-CSW: text steganalysis and hidden capacity estimation based on convolutional sliding windows. Multimed. Tools Appl. 1–24 (2020)
31. Yang, Z., Wang, K., Ma, S., Huang, Y., Kang, X., Zhao, X.: IStego100K: large-scale image steganalysis dataset. In: Wang, H., Zhao, X., Shi, Y., Kim, H.J., Piva, A. (eds.) IWDW 2019. LNCS, vol. 12022, pp. 352–364. Springer, Cham (2020). https://doi.org/10.1007/978-3-030-43575-2_29
32. Yang, Z., Wei, N., Liu, Q., Huang, Y., Zhang, Y.: GAN-TStega: text steganography based on generative adversarial networks. In: Wang, H., Zhao, X., Shi, Y., Kim, H.J., Piva, A. (eds.) IWDW 2019. LNCS, vol. 12022, pp. 18–31. Springer, Cham (2020). https://doi.org/10.1007/978-3-030-43575-2_2
33. Yang, Z., Wei, N., Sheng, J., Huang, Y., Zhang, Y.J.: TS-CNN: text steganalysis from semantic space based on convolutional neural network. arXiv preprint arXiv:1810.08136 (2018)
34. Yang, Z., Zhang, P., Jiang, M., Huang, Y., Zhang, Y.-J.: RITS: real-time interactive text steganography based on automatic dialogue model. In: Sun, X., Pan, Z., Bertino, E. (eds.) ICCCS 2018. LNCS, vol. 11065, pp. 253–264. Springer, Cham (2018). https://doi.org/10.1007/978-3-030-00012-7_24
35. Yang, Z., Zhang, S., Hu, Y., Hu, Z., Huang, Y.: VAE-stega: linguistic steganography based on variational auto-encoder. IEEE Trans. Inf. Forensics Secur. **16**, 880–895 (2020)
36. Ziegler, Z.M., Deng, Y., Rush, A.M.: Neural linguistic steganography. arXiv preprint arXiv:1909.01496 (2019)

On the Sharing-Based Model of Steganography

Xianfeng Zhao[1,2](✉) ⓘ, Chunfang Yang[3], and Fenlin Liu[3]

[1] State Key Laboratory of Information Security, Institute of Information Engineering, Chinese Academy of Sciences, Beijing 100195, China
zhaoxianfeng@iie.ac.cn
[2] School of Cyber Security, University of Chinese Academy of Sciences, Beijing 100093, China
[3] State Key Laboratory of Mathematical Engineering and Advanced Computing, Zhengzhou, China
chunfangyang@126.com, liufenlin@vip.sina.com

Abstract. Steganography has long been considered as a way of hiding the fact of secret communication. However, the fact that a message sender and a receiver communicated with each other is seldom protected. It can be dangerous because they often have very secret relationship. Noticeably, in recent two years some researchers respectively proposed robust and secure steganographic schemes which realize dependable steganography in the lossy channel of social media without degrading the security and capacity, and more relevant technologies, such as robust steganographic coding and fast steganography, were also proposed. In this review, the sharing-based model of steganography implied by social media and promoted by both the emerging advanced robust steganography and fast steganography is formally presented and compared with the dominating prisoner-warden model. It is shown that designing steganography under the sharing-based model can protect the fact of communication and the sender-receiver relationship instead of merely the fact of secret communication, improving the overall security of steganographic systems.

Keywords: Steganography · Steganalysis · Security model · Social media · Covert communication

1 Introduction

Steganography is an important way of covert communication in which a sender hides secret message into cover content and sent the ostensibly natural stego-content to a receiver. In contrast to cryptography, it protects the fact of secret activity instead of only the content of secret message. The use of steganography has a long history but it had not been scientifically developed until the last decade of the 20th century, when people begun to use computer and multimedia in their daily life. Modern steganography often takes multimedia, such as image, video, and audio, as covers because they have more redundant information.

To define the roles and the scenarios in steganography, Simmons [1] proposed the famous *prisoner-warden model*. In the model, which is illustrated in Fig. 1, two prisoners Alice and Bob want to exchange secret message but can only ask the warden

© Springer Nature Switzerland AG 2021
X. Zhao et al. (Eds.): IWDW 2020, LNCS 12617, pp. 94–105, 2021.
https://doi.org/10.1007/978-3-030-69449-4_8

Eve to deliver stego-notes. In fact, Alice and Bob act as a sender and a receiver respectively in covert communication, and Eve means a monitored channel. Alice and Bob need conceal the fact of secret activity such as encryption, and Eve want to detect it. If Eve modifies the notes, the channel is lossy; otherwise, it is lossless. Since most multimedia delivery over network is point-to-point and lossless before the wide use of social media, the prisoner-warden model using non-robust steganography has dominated the research. In this direction, a series of algorithms and methods, together with the analysis of them, have been proposed. This kind of steganography is designed to suppress the embedding perturbation under a payload of secret message and the assumption of no channel noise. The most important principles and methods of them are collected in [2] and [3].

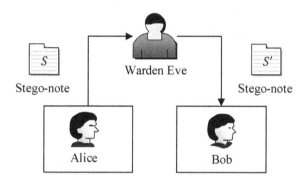

Fig. 1. Illustration of the prisoner-warden model. S' may be equal to S or not. The former case means a lossless channel, and the latter a lossy channel.

However, in fact the prisoner-warden model can be weak if a sender and a receiver have secret relationship which needs protected. And unfortunately, it is often the case in the real world. Intuitively, sharing stego-content over social media can alleviate the problem because the content is sent in a multicast way so that the expected receiver is protected. But as we know, the multimedia channels of social media are often designed lossy for saving bandwidth and storage space. Transcoding is widely used in these channels to compress the media and decrease the resolution sizes. As a result, such protection can be impossible if only the traditional steganography is used.

To fulfill the demand of robust steganography which can make use of social media channel by resisting the lossy processing, some new schemes and techniques have been proposed [4–15]. An important part of them [4–7], which we call the advanced robust steganographic schemes, realize the needed robustness without degrading the security and capacity. Moreover, the development of fast steganography [16–18] is enabling running the sophisticated steganographic software smoothly on mobile devices, which are used as the primary way of accessing social media. In this review, we indicate that the significance of them is not restricted to only exploiting new channels and new platforms. Based on applying them to the content sharing channels over social media, we define the sharing-based model of steganography and discuss its advantages. And then we will introduce some technologies which support applying the model over social media.

2   Limitation of the Prisoner-Warden Model

Steganographers need hide their particular communication behavior and steganalysts want to detect it. For the design of steganography, it is important to clarify the meaning of such behavior and to what extent and scope it should be protected. Because steganography is usually utilized in hostile environments, it is more likely that steganographers want to conceal not only the fact of secret communication but also the fact of communication. That is, the relationship between a sender and a receiver also needs protected so that they can avoid incriminating each other.

Under the above requirement, the dominating prisoner-warden model shows limitation. Alice and Bob only protect the fact of secret communication by asking Eve to pass stego-notes. In fact, it assumes that contents can only be delivered in a point-to-point way. It was reasonable at the time when multicast or broadcast delivery of contents had not been widely used. As a result, Eve will also accuse Bob if he has detected stego-notes from Alice, and vice versa. It makes a steganographic scheme much weak in security as a system. In the time of big data analysis, one can assume a monitor knows all contact information. Consequently, the sender-receiver relationships in steganographic systems designed under the prisoner-warden model are unprotected. It becomes more and more unreasonable because nowadays people widely share contents in multicast or broadcast ways.

3   Sharing-Based Model Implied by Social Media

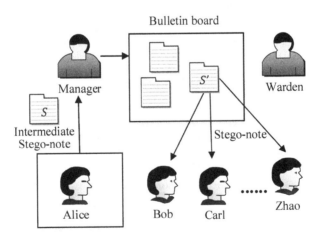

Fig. 2. Illustration of the sharing-based model, in which Alice and Bob hide the fact of communication and their sender-receiver relationship.

Fig. 3. The wide use of sharing content over social media implies the sharing-based model but how to make steganography robust and secure remains as a challenge. *S* is called an intermediate medium.

To strengthen the security, a *sharing-based model*, illustrated in Fig. 2, can be defined from the above prisoner-warden model. Suppose that a bulletin board is shared among all prisoners. Each one can put notes on it. Then, Alice and Bob can exchange message by putting stego-notes there. It means using multicast channel. The bulletin board manager may or may not revise the notes. Similarly, the former case means a lossy channel, and the latter a lossless one. Because all prisoners may read the notes, the warden has difficulty to recognize the sender-receiver relationship between Alice and Bob. Interestingly, it seems that the sharing-based model had been used before the emergence of Internet. For examples, in some films or stories, early spies had their stego-notes posted on newspapers so that their comrade could receive secret message without revealing their relationship. Similarly, some stego-notes could be broadcasted by radio. They can be regarded as the early application of sharing-based model in which protecting contact relations is extremely important.

In recent two decades, social media has provided people with a more and more ubiquitous and feasible way of sharing content (see Fig. 3), implying that the sharing-based model becomes more applicable. For examples, people can share contents on blogs in a broadcast way, and they can choose to share them at a virtual circle of friends in a multicast way. Noticeably, it was reported that in 2010 [19, 20], when lossy processing was seldom used over network, some Russian spies arrested in the United States hided messages in online public images to conceal the contact relation. However, we think that the early effort to utilize the sharing-based model over social media was hindered because such networks begun to apply transcoding to the delivered content for saving bandwidth and storage space. Transcoding often consists of both lossy compression and scaling down the resolution size of multimedia (see Fig. 4). As a result, the traditional steganography becomes inapplicable. Such processing may happen at the server or at the mobile front end. In either case, it is hard to avoid being transcoded by breaking the system.

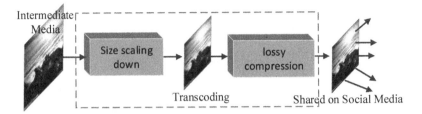

Fig. 4. Transcoding often consists of both lossy compression and scaling down the resolution size of multimedia.

In the next section, we will show that the obstacle to applying steganography and the sharing-based model over social media is being removed by the development of some supporting techniques such as the advanced robust steganography and fast steganography. Of course, in some social media systems, a user is allowed to particularly set the system so as to send out multimedia in a lossless way. However, in most cases it only applies to point-to-point communication and such operation is seldom used by a normal user. Because a steganographer tries to reveal normal behavior over networks, such treatment violates his or her principles.

4 Techniques Which Support Sharing-Based Model

4.1 Technical Requirement of Applying the Model

To apply the sharing-based steganography over social media, we think three kinds of technologies are needed.

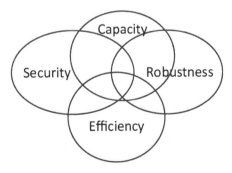

Fig. 5. The challenge of applying sharing-based model over social media is to coordinate some performances that seemingly conflicts with each other.

Channel Modeling, Simulation, and Exploitation. Before any new design of robust and secure steganography, the lossy processing ways of a channel should be recognized and the parameters estimated. Then the channel can be simulated, tested locally and exploited online.

Robust and Secure Embedding with Capacity. New steganography should not only have enough robustness to ensure reliable communization but also keep the covertness and capacity as more as possible.

Fast and Secure Embedding with Capacity. New steganography should adapt to the mobile devices so that it must be efficient. We have said that a steganographer tries to reveal normal behavior over networks. So, if most people use mobile phone to send contents, he or she should do it likewise.

By the requirement, the challenge of applying sharing-based model over social media is to coordinate the seemingly conflicting performances, including security, robustness, efficiency, and capacity (see Fig. 5). In the followed sub-sections, we will introduce some of the supporting technologies and point out that the knack of fulfilling such a tough task is to make the steganography adapt to the channels. Some techniques for designing the robust steganography and fast steganography have been developed for years but they may not directly aim at applying the sharing-based model.

4.2   Channel Modeling, Simulation, and Exploitation

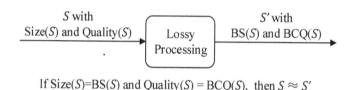

If Size(S)=BS(S) and Quality(S) = BCQ(S), then $S \approx S'$

Fig. 6. Modeling of the lossy multimedia processing made by some typical social media networks, in which BS is for borderline size, and BCQ is for borderline compression quality.

Table 1. Some typical parameters of uploaded image and those of the downloaded, where the boldfaced ones can be estimated as borderline parameters.

Format	Upload size	Upload QF	Download size	Download QF
JPEG	3000 × 2000	95	2048 × 1365	85
		85	**2048 × 1365**	**85**
		65	2048 × 1365	85
	2048 × 1365	95	2048 × 1365	85
		85	**2048 × 1365**	**85**
		75	2048 × 1365	71
		71	**2048 × 1365**	**71**
		65	**2048 × 1365**	**65**
	512 × 512	95	512 × 512	85
		71	512 × 512	71
		65	512 × 512	65

Apparently, to apply sharing-based model over social media depends on successful design of the advanced robust steganography, which not only resists the attack of transcoding but also keeps as more as security and capacity. An important issue here is how to recognize the channel's processing and estimate the parameters so that new steganography can be designed accordingly.

A typical method for channel recognition and estimation has been briefly introduced in [4]. Here, we introduce and formalize it as follows. First, media samples are uploaded onto the social media. Then, the transcoded samples can be downloaded and compared with the uploaded ones. Often the processing, including size scaling and content recompression, is easy to recognize and the parameters can be estimated through the comparison. Zhao *et al.* [4] observed that most social media have two kinds of following parameter as illustrated in Fig. 6 and Table 1, which is extremely important for designing the advanced robust steganography.

Borderline Size (BS) Parameter. It is a size to which the channel changes an uploaded medium. And the channel does not further decrease the size of an uploaded medium if it is equal to or less than the BS. The BSs turn to be larger with the improvement of computing capability. For example, 2048×1365 is a typical BS for images as shown in Table 1.

Borderline Compression Quality (BCQ) Parameter. It is a compression quality to which the channel changes an uploaded medium. And the channel does not further change the quality of an uploaded medium if it is equal to or less than the BCQ, though a recompression with the same parameters may be applied. Similarly, the BCQs turn to be higher with the improvement of computing capability. For example, quality factor (QF) 85 is a typical BCQ for images as shown in Table 1.

Consequently, the existence of BS parameter and BCQ parameter enables steganographers to degrade the channel attack into a simpler one. Since designing a scheme robust to size scaling is very difficult, some relevant robust steganographic schemes try to avoid the scaling by using covers with BS as their sizes. In these cases, the attack degrades to a recompression attack. If a delivered content with BS as its size has a quality equal to BCQ, the attack is further degraded to a recompression with the same quality parameters. The degraded attacks introduce less noises so that the embedding energy can be smaller to keep the enough security.

4.3   Robust and Secure Steganography

By making use of the borderline parameters, current design of robust schemes can only consider resisting lossy compression under the assumption that only borderline sized media are used as covers. The difference is only whether BCQ is also used as the cover quality. Up to now, many efforts [4–15] have been made to design robust and secure image and video steganography, though some of them might only aim at exploiting new point-to-point channels at the time.

Zhang *et al.* [8–12] proposed some watermarking-like techniques for designing a class of robust steganography. The techniques, including coefficient inequality relation adjustment, region selection, and adaptive dither modulation, have been respectively or jointly used to ensure reliable robustness. The level of robustness can be adjusted according to the channel by changing the strength of the inequality or some relevant parameters. Typically, in the DCT coefficient relationship based adaptive steganography (DCRAS) [12], the embedded 0 and 1 are respectively represented by two inequalities of same-mode JPEG coefficients in neighboring 4 blocks. The embedding can adjust one of the coefficient values to satisfy either of them to embed 1 or 0. The strength of the inequalities can be adjusted by adapting an additive parameter of the inequality to the channel. The larger the parameter, the more the robustness needed. The parameter value is estimated by simulating the channel locally. After the basic embedding way is decided, distortions can be computed and then syndrome trellis codes (STC) [21] can be used. In [10] featured regions are selected for improve the robustness, and in [9], adaptive dither modulation, which takes quantization tables for adaptively deciding the step sizes, is also used. However, the schemes give much lower capacity. Only when the capacity is from 10% to 30% of a normal level, they have accepted security. A detailed performance evaluation of this class of robust steganography has been reported in [22].

To acquire the robustness, new ways of applying steganographic codes and error correction codes (ECC) have also been proposed. Kin-Cleaves and Ker [13] designed the dual STC by combining both the correction treatment and the embedding in STC processing. A stego-image is made to have two syndromes simultaneously, and one of them is used for error correction in supporting the robust extraction of message. Experimental results show that it outperforms a straightforward combination of the standard Reed Solomon (RS) codes and STC. Based on the observation that correcting stego-content is more effective than correcting message in mitigating the error propagation of STC decoding, Feng et al. [14] proposed to encode the output of STC processing and correct stego-content before message extraction. To resist the continuous errors of STC decoding, Bao et al. [15] proposed to use concatenated ECC encoder with interweaving between the two sub-encoders. In decoding, the interweaving between two sub-decoders disperses the continuous error resulting from STC decoding so that the outer sub-decoder becomes more effective. These methods help improve the robustness but often they cannot get rid of bit error independently.

More advanced robust steganography should have the robustness just enough to resist the channel attack and maintain as more as security and capacity. Such a steganographic scheme must find ways of exploiting the channel more accurately. As we have discussed, if a delivered content has the size equal to BS and compression quality equal to BCQ, the attacking noise, which only results from the recompression under the same parameter, has the minimal energy. Moreover, Zhao et al. [4] observed that when a JPEG image has been compressed under one QF many times, it becomes stable, i.e., compressing it again almost changes no JPEG efficient of the image. This result is due to the fact that two noises introduced by the same-parameterized JPEG recompression, i.e., spatial quantization noise and coefficient quantization noise, turn to diminish with each other. So, in the channel matching-based scheme, the embedding and compression under the channel QF are repeated locally to let the intermediate stego-image stable so that the lossy channel cannot affect it much. Tao et al. [5] also proposed a channel matching-based scheme, in which a cover image is compressed locally as in the channel and then embedded. The resulting stego-image serves as a reference to the robust adjustment of the cover image. However, this design assumes that there is no quantization noise introduced by the pixel value rounding in spatial domain, and it is not often the real case. In the scheme proposed by Lu *et al.* [6], how to simulate and adjust the intermediate stego-image is learnt and executed by a neural network, called auto-encoder. Consequently, more kinds of channels, which may have some particular processing, can be automatically simulated. The above channel matching-based methods, with the aid of ECC, can have a zero or very low bit error rate with the capacity and security at the same level of their non-robust counterparts. Although in [5] an adjusted intermediate image before lossy processing is more detectable, in many cases lossy processing occurs at mobile devices so that steganalysts cannot have the change to detect it.

More recently, Fan *et al.* [7] proposed a robust video steganographic scheme, in which side information is embedded into another standalone domain to record the selected frames where better performance of robustness has been given in local testing. It has been shown that at a very little price of capacity and similarly with the aid of ECC, a bit error rate less than 1% can be achieved without degrading the security. In some cases, the error rate is 0%. In fact, such idea can be expanded to the design of many side information-based robust steganographic schemes if they take both a data carrying domain and a side information carrying domain in a single multimedium. For examples, the accompanying voice can be used to record the robust video places, and the different video domains and macroblocks, which do not interfere with each other, can be used as the side information channel and the communication channel respectively.

4.4 Fast and Secure Steganography

Because most people use mobile devices to access social media, in the long run another requirement for applying steganography and sharing-based model over social media is to adapt the steganographic technique to these devices. In general, a mobile device has much less computation power compared with a personal computer, and the dominating adaptive steganography, such as the typical scheme named JUNIWARD (JPEG UNIversal WAvelet Relative Distortion) [23], needs more computing resources than common applications to improve the security. As a result, the challenge to fulfil the above requirement is to design fast and secure steganography. Some pioneer work [16–18] has designed fast steganography at very small price of security and capacity. We shall briefly review them as follows.

UED (Uniform Embedding Distortion) [24] has long been regarded as a fast steganographic scheme which directly evaluates embedding distortion in the embedding coefficient domain. Nevertheless, its security under steganalysis is inferior to the more secure JUNIWARD which evaluates the distortion in wavelet domain. To improve the security of UED but still maintain the computational efficiency, Guo *et al.* [16] proposed the UERD (Uniform Embedding Revisited Distortion). The scheme jointly evaluates the distortion for modifying a coefficient with both the coefficient's quantization step value and a value indicating the sub-block's complexity. The efficiency is built on the following 2 facts. First, the coefficient's quantization step value is constant, and second, a sub-block's complexity only needs computed once for all coefficients in it. Experimental results [18] show that the time for computing the distortion of UERD is about 1/6 of the JUNIWARD, and the security level of UERD under steganalysis is almost the same as that of JUNIWARD.

Besides the time for computing the distortion, another part of running time of adaptive steganography is the time for computing STC. In general, adaptive steganography takes all distortions, that is, distortions computed on every coefficient or pixel, into the STC computation. Even if one only need embed in a low capacity, he or she has to process all the image. Li *et al.* [17] proposed some ways of speeding up STC by shortening the length of input. In their method, the cover coefficients can be selected according to the values of their JPEG quantization steps. As we know, the smaller a quantization step, the smaller the modified amplitude and the distortion value. So, only coefficients with smaller JPEG quantization steps are selected and embedded. Another scheme is based on segmentation of a cover. The new length of input is the original length divided by the shortened times. Suppose the times is two. Each two neighboring 4-element segment is summed. On the same places in each segment, distortion is the smallest one. STC is applied to the summed sequence so that the path is shortened. And it has been shown that under such treatment messages can be extracted correctly. Experimental results show that the detection error rate difference between the tradi- tional use of STC and this scheme is no more than 1% under the steganalytic scheme named DCTR (Discrete Cosine Transform Residual) [25].

Su *et al.* [18] proposed a fast steganographic scheme by simplifying the distortion computation of JUNIWARD and concurrently computing the segmented STC paths. JUNIWARD computes the embedding distortion by evaluating the change of wavelet coefficients in the HL, LH, and HH sub-bands. However, the coefficients' values in the 3 sub-bands have symmetry so that two of them is omitted in the scheme named fast JUNIWARD in [18], saving about 2/3 of the time for computing the distortions without degrading the security and capacity. In addition, experimental results show that if STC paths, whose inputs are the permutated embedding domain and distortions, are seg- mented and run separately and concurrently, the security under steganalysis will only be affected very slightly. In the fast JUNIWARD, the paths are often segmented into 4 parts which are run concurrently in different CPU kernels in a mobile device, and the error rate of it under DCTR steganalysis decreases by only about 1%–2% under the payload of 0.3 bpnzAC (bit per non-zero Alternating-current Coefficient).

5   Concluding Remarks and Future Work

With the development of advanced robust steganography and fast steganography over social media, the sharing-based model, which not only aims at protecting the secret activity but also aims at protecting the sender-receiver relationship, becomes more applicable and appealing. We think that it already has and will keep having remarkable influence on the evolvement of steganographic technologies. More schemes are to be designed robust at little price of capacity and security. And designing fast and secure steganography fit for mobile devices and combining them with the robust steganog- raphy become extremely important for better exploiting social media networks. Moreover, how to build the upload and download channel in a secure manner will be also important. Finally, it is interesting that the techniques of robustness and the techniques of covertness, previously for watermarking and steganography respectively, come together in this scenario. We can expect new novel methods that combine and balance them more cleverly and properly.

Acknowledgements. This work was partly supported by National Natural Science Foundation of China (NSFC) under the grant numbers U1736214, 61972390, 61902391, 61802393, and 61872356.

References

1. Simmons, G.J.: The prisoner's problem and the subliminal channel. In: Chaum, D. (ed.) Proceedings of the CRYPTO 1983, Santa Barbara, CA, USA, 21–24 August 1983, pp. 51–67. Plenum Press, New York (1984)
2. Fridrich, J.: Steganography in Digital Media: Principles, Algorithms, and Applications. Cambridge University Press, Cambridge (2010)
3. Zhao, X., Zhang, H.: Steganology: Principles and Technologies. Science Press, Beijing (2018). (in Chinese)
4. Zhao, Z., Guan, Q., Zhang, H., Zhao, X.: Improving the robustness of adaptive steganographic algorithms based on transport channel matching. IEEE Trans. Inf. Forensics Secur. **14**(7), 1843–1856 (2019)
5. Tao, J., Li, S., Zhang, X., Wang, Z.: Towards robust image steganography. IEEE Trans. Circuits Syst. Video Technol. **29**(2), 594–600 (2019)
6. Lu, W., Zhang, J., Zhao, X., Zhang, W., Huang, J.: Secure robust JPEG steganography based on autoencoder with adaptive BCH encoding. IEEE Trans. Circuits Syst. Video Technol. (2020). https://doi.org/10.1109/TCSVT.2020.3027843
7. Fan, P., Zhang, H., Cai, Y., Xie, P., Zhao, X.: A robust video steganographic method against social networking transcoding based on steganographic side channel. In: Amerini, I., Bestagini, P., Pevny, T. (eds.) Proceedings of the IH & MMSec 2020, Denver, CO, USA, 22–24 June 2020, pp. 127–137. ACM Press, New York (2020)
8. Zhang, Y., Luo, X., Guo, Y., Qin, C., Liu, F.: Multiple robustness enhancements for image adaptive steganography in lossy channels. IEEE Trans. Circuits Syst. Video Technol. **30**(8), 2750–2764 (2020)
9. Zhang, Y., Zhu, X., Qin, C., Yang, C., Luo, X.: Dither modulation based adaptive steganography resisting JPEG compression and statistic detection. Multimed. Tools Appl. **77**, 17913–17935 (2018)
10. Zhang, Y., Luo, X., Yang, C., Liu, F.: Joint JPEG compression and detection resistant performance enhancement for adaptive steganography using feature regions selection. Multimed. Tools Appl. **76**(3), 3649–3668 (2016). https://doi.org/10.1007/s11042-016-3914-0
11. Zhang, Y., Luo, X., Yang, C., Ye, D., Liu, F.: A framework of adaptive steganography resisting JPEG compression and detection. Secur. Commun. Netw. **9**(15), 2957–2971 (2016)
12. Zhang, Y., Luo, X., Yang, C., Ye, D., Liu, F.: A JPEG-compression resistant adaptive steganography based on relative relationship between DCT coefficients. In: Weippl, E. (ed.) Proceedings of the 10th International Conference on Availability, Reliability and Security, Toulouse, France, 24–27 August 2015, pp. 461–466. IEEE Press, Piscataway (2015)
13. Kin-Cleaves, C., Ker, A.D.: Adaptive steganography in the noisy channel with dual-syndrome trellis codes. In: Yuen, P.C., Huang, J. (eds.) Proceedings of the IEEE WIFS 2018, Hong Kong, China, 10–13 December 2018, pp. 1–7. IEEE Press, Piscataway (2018)

14. Feng, B., Liu, Z., Wu, X., Lin, Y.: Robust syndrome-trellis codes for fault-tolerant steganography. In: Jain, L.C., Peng, S.-L., Wang, S.-J. (eds.) SICBS 2019. AISC, vol. 1145, pp. 115–127. Springer, Cham (2020). https://doi.org/10.1007/978-3-030-46828-6_11

15. Bao, Z., Guo, Y., Li, X., Zhang, Y., Xu, M., Luo, X.: A robust image steganography based on the concatenated error correction encoder and discrete cosine transform coefficients. J. Ambient Intell. Hum. Comput. **11**, 1889–1901 (2019). https://doi.org/10.1007/s12652-019-01345-8

16. Guo, L., Ni, J., Su, W., Tang, C., Shi, Y.-Q.: Using statistical image model for JPEG steganography: uniform embedding revisited. IEEE Trans. Inf. Forensics Secur. **10**(12), 2669–2680 (2015)

17. Li, W., Zhou, W., Zhang, W., Qin, C., Hu, H., Yu, N.: Shortening the cover for fast JPEG steganography. IEEE Trans. Circuits Syst. Video Technol. **30**(6), 1745–1757 (2020)

18. Su, A., Ma, S., Zhao, X.: Fast and secure steganography based on J-UNIWARD. IEEE Signal Process. Lett. **27**, 221–225 (2020)

19. Calamia, J.: How the Russian spies hid secret messages in public, online pictures, Discover, 2 July 2010. https://www.discovermagazine.com/technology/how-the-russian-spies-hid-secret-messages-in-public-online-pictures. Accessed 17 Oct 2020

20. Stier, C.: Russian spy ring hid secret messages on the web, NewScientist, 2 July 2010. https://www.newscientist.com/article/dn19126-russian-spy-ring-hid-secret-messages-on-the-web/. Accessed 17 Oct 2020

21. Filler, T., Judas, J., Fridrich, J.: Minimizing additive distortion in steganography using syndrome-trellis codes. IEEE Trans. Inf. Forensics Secur. **6**(3), 920–935 (2011)

22. Zhang, Y., Qin, C., Zhang, W., Liu, F., Luo, X.: On the fault-tolerant performance for a class of robust image steganography. Sig. Process. **146**(2018), 99–111 (2018)

23. Holub, V., Fridrich, J., Denemark, T.: Universal distortion function for steganography in an arbitrary domain. EURASIP J. Inf. Secur. **2014**, Article number 1, 1–13 (2014)

24. Guo, L., Ni, J., Shi, Y.-Q.: An efficient JPEG steganographic scheme using uniform embedding. In: Campisi, P., Kundur, D. (eds.) Proceedings of the IEEE WIFS 2012, Tenerife, Spain, 2–5 December 2012, pp. 169–174. IEEE Press, Piscataway (2012)

25. Holub, V., Fridrich, J.: Low complexity features for JPEG steganalysis using undecimated DCT. IEEE Trans. Inf. Forensics Secur. **10**(2), 219–228 (2015)

Watermarking

Towards Informed Watermarking of Personal Health Sensor Data for Data Leakage Detection

Sebastian Gruber[1(✉)], Bernd Neumayr[1], Christoph Fabianek[2],
Eduard Gringinger[2], Christoph Georg Schuetz[1], and Michael Schrefl[1]

[1] Johannes Kepler University Linz, Linz, Austria
{sebastian.gruber,bernd.neumayr}@jku.at
[2] OwnYourData, Bad Vöslau, Austria

Abstract. Users of personal health devices want an easy way to permanently store their personal health sensor data and to share them with physicians and other authorized users, trusting that the data will not be disclosed to third parties. Digital watermarking for data leakage detection aims to prevent the unauthorized disclosure of data by imperceptibly marking the data for each authorized user, so that the authorized user can be identified as the data leaker and be held accountable. In this paper we present an approach for digital watermarking conceived as part of a personal health sensor data management platform. The approach comprises techniques for informed watermark embedding and non-blind watermark detection. Based on a proof-of-concept prototype, the approach is evaluated regarding configurability, robustness, and performance.

Keywords: Medical sensor data · Digital fingerprinting · Time series data

1 Introduction

Personal health sensor data are recorded by wearable devices that keep track of different indicators of an individual's health. A diabetes patient, for example, may keep with them a sensor for measuring blood glucose levels. The constant monitoring of such sensitive personal information puts users of personal health sensors in a dilemma. On the one hand, users of personal health devices want to retain control over the data collected using those sensors. On the other hand, in order for the collected data to be useful in preventing health issues, users want an easy-to-use platform to manage and permanently store these data, and possibly share these data with selected other parties.

A management platform for personal health sensor data should facilitate sharing the data with other individuals, e.g., physicians, or applications, e.g., for the analysis of data in the context of pharmaceutical research. In order to retain control over the data, first and foremost, data access should only be given to authorized data users. Furthermore, every request for access to personal data should be logged to monitor and audit requests for data by authorized users. In this regard, in case of data leakage, it should be possible to identify and hold accountable the authorized user who leaked the data. To this end, digital watermarking can be employed.

© Springer Nature Switzerland AG 2021
X. Zhao et al. (Eds.): IWDW 2020, LNCS 12617, pp. 109–124, 2021.
https://doi.org/10.1007/978-3-030-69449-4_9

In this paper, we propose an approach for the provisioning of watermarked personal health sensor data and for the detection of data leakage. The approach is configurable to balance the trade-off between usability of watermarked data and robustness against attacks. The approach is evaluated using a proof-of-concept prototype with respect to configurability, performance, and robustness against rounding value, random value, deletion, subset selection, and mean collusion attacks.

The remainder of the paper is structured as follows. Section 2 briefly introduces the data that should be watermarked and sketches the overall process. Section 3 describes the approach in detail. Section 4 evaluates the approach based on a proof-of-concept prototype. Section 5 describes a specific application scenario and the deployment of the approach. Section 6 points to related work. Section 7 concludes the paper.

2 Problem Analysis and Overall Process

In this section we briefly introduce the data that should be watermarked as well as the user and data access structure of a personal health sensor data platform. We sketch an overall process for provisioning watermarked personal health sensor data and for data leakage detection.

Let us first look at the data that are subject to watermarking. Every health sensor produces a time-indexed sequence of measurements, typically with uniform intervals, e.g. one measurement every five minutes. Every measurement comprises the device id, the type of measurement, and the unit of measurement – these first three attributes are fixed for each sensor – as well as the date, time, and the measured value. When grouping the data by sensor and date, a sequence of time-value pairs remains. Such a group of data is what we refer to as a *dataset fragment* (see Fig. 1 for an example), or simply a *fragment*.

```
{ "deviceId": "D32133",            { "deviceId": "D32133",
  "type": "cbg",                     "type": "cbg",
  "unit": "mmol/L",                  "unit": "mmol/L",
  "time": "2017-02-04T00:02:23"      "day": "2017-02-04"
  "value": 6.993942 },               "measurements": [
{ "deviceId": "D32133",               { "time": "00:02:23",
  "type": "cbg",                        "value": 6.993942 },
  "unit": "mmol/L",                   { "time": "00:07:23",
  "time": "2017-02-04T00:07:23"         "value": 7.923523 },
  "value": 7.923523   },              { "time": "00:12:23",
...                                     "value": 6.434981 },
{ "deviceId": "D32133",               ...
  "type": "cbg",                      { "time": "23:57:23",
  "unit": "mmol/L",                     "value": 12.02332 },
  "time": "2017-02-04T23:57:23"       ]
  "value": 12.02332 },              }
```

Fig. 1. Example sensor data (left) and its grouping into a fragment (right) in JSON format

In our personal health sensor data management platform, a dataset fragment is treated as an atomic, i.e., basic and indivisible, unit for data sharing and watermarking. The health sensor data pool, i.e., the data store of the platform, is regarded as a set of such fragments.

In the following, we briefly introduce our personal health sensor data platform's user, authorization and access structure. We also sketch the overall process of provisioning watermarked person health sensor data; details are discussed in Sect. 3. Note that the only purpose of watermarking in our setting is to find out, given leaked data, who, out of a set of authorized users, had originally requested (accessed) the leaked data on the platform and subsequently disclosed the data.

We assume that there are many individuals who store their personal health sensor data on the platform and many authorized data users (persons and applications who request data) but only a few authorized data users per dataset fragment. Further, only users who have actually requested (accessed) some fragment are potential leakers of that fragment.

A central design choice concerns multiple requests by the same data user of the same or overlapping data: If a user requests the same fragment multiple times, even in different requests (e.g. once all blood glucose data of yesterday for all persons and once the blood glucose data of Person X for the last week), then this fragment (e.g., yesterday's blood glucose data of Person X) should always be watermarked the same way. Otherwise, a single user could perform many similar requests and, with many differently watermarked versions of the same fragment, could be able to effectively remove the watermark.

When a data user issues a request for data, the request is first broken down into multiple requests, each answered by a single dataset fragment. When a user requests a fragment for the first time, they are permanently associated with a watermark generated for that fragment. Every time that same user now requests that particular fragment, the same watermark associated with the combination of user and fragment is embedded into the returned fragment.

Data leakage detection is initiated when a suspicious dataset is found in order to answer two questions. First, does the dataset originate from the health sensor data pool? Second, which of the authorized users leaked the data? These questions are not directly answered for the suspicious dataset as a whole, but the suspicious dataset is partitioned into dataset fragments and, by using non-blind watermark detection, these questions are answered per fragment before being combined into an overall answer. Answering these questions is complicated by the possibility that data leakers may have tried to remove the watermark in the case of which we speak of noisy watermarked data.

3 Design

In this section we introduce the design of the watermarking approach. Watermark embedding is used when a data user requests a dataset. Being aware of the data being watermarked, malicious data users could attack the datasets to damage or remove the watermarks before leaking the data. If a suspicious dataset is found, watermark detection is used to identify the potential data leakers. Because of possible attacks, the watermark detection must take modifications of datasets and their watermarks into account.

3.1 Preliminaries

Concerning watermark generation/embedding we make the following design choices. (i) Watermarks are embedded only in the value field of measurements. With the fixed time intervals, introducing errors to the time fields would be easily perceptible. (ii) A watermark introduces a small error to every measurement value, and not only to selected measurement values; this increases robustness against collusion attacks, deletion attacks, and subset selection attacks without hampering usability. (iii) Watermark generation is informed: by considering the original, errors can be introduced without making the watermark perceptible. (iv) Watermark generation is configurable by usability constraints associated with type and unit of measurement to balance the trade-off between usability of watermarked data and robustness against attacks. (v) A secret key associated with each fragment keeps the watermark generation secure even if the algorithm is publicly known (see [1]). (vi) To make watermarks reproducible and avoid the necessity of storing the generated watermarks in the database, their generation is deterministic based on the original fragment, usability constraint, the fragment's secret key, and a watermark number associated in the request log with a user's (possibly repeated) requests of a fragment.

We realized the usability constraint as a concept consisting of the maximum error, the sensor's minimum and maximum value as well as the number of ranges. The maximum error determines the (initial) maximum error range for each measurement value. The maximum possible error range of each measurement may then be limited by the sensor's minimum and maximum value. The error must not fall below the minimum value or exceed the maximum value. Finally, the number of ranges configures the error distribution by indicating the number of error sub-ranges within the possible error range with each error sub-range having a probability with which it is selected. The error sub-ranges are halved into error sub-ranges below and above the original value. The closer an error sub-range to the original value, the higher the probability to be selected. In other words, the higher the number of ranges, the closer the errors are distributed to the original value. Table 1 provides some example selection probabilities of error sub-ranges using various configurations regarding the number of ranges. In summary, a tight usability constraint, i.e., small maximum error and high number of ranges, results in small errors providing high imperceptibility but less security.

Concerning watermark detection we make the following design choices. (i) Similarity search is used to compute the probability of a suspicious fragment being a fragment from the database. In case an attacker modifies a fragment by any means, similarity search provides the possibility to identify the matching fragment from the database. An increasing database size, however, requires an efficient similarity search which is out of scope. (ii) Similarity search is also used to compute the probability of an extracted watermark being an already embedded watermark. In case an attacker corrupts a watermark by any means, similarity search provides the possibility to identify the matching watermark. The downfall of using similarity search is that also innocent data users are regarded as potential data leakers.

Table 1. Example selection probabilities of error sub-ranges using different number of ranges

Number of ranges	Selection probabilities of error sub-ranges of the lower error range				Selection probabilities of error sub-ranges of the upper error range				Total
	4	3	2	1	1	2	3	4	
2				50%	50%				100%
4			25%	25%	25%	25%			100%
6		12.5%	12.5%	25%	25%	12.5%	12.5%		100%
8	6.25%	6.25%	12.5%	25%	25%	12.5%	6.25%	6.25%	100%

3.2 Watermark Embedding

Whenever an authorized data user requests data – which corresponds to a set of requested fragments –, watermark embedding is performed. For each requested fragment, a matching watermark associated with its requesting data user is generated and embedded resulting in a watermarked fragment. The combination of all watermarked fragments is transmitted to the requesting data user as a watermarked dataset that fulfils the user's request. The approach for watermark embedding is illustrated in Fig. 2 and discussed in more details below.

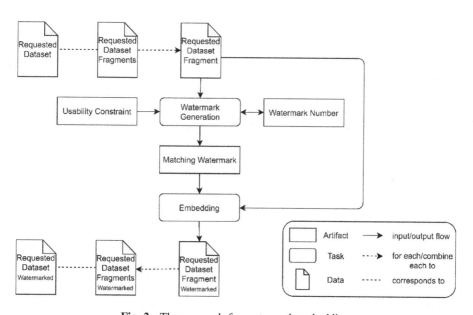

Fig. 2. The approach for watermark embedding

Task 1 (Watermark Generation). Watermark generation generates a matching watermark for each requested fragment based on the requested fragment, a usability constraint, and existing requests.

The following steps are performed for each requested fragment. (1) The requested fragment's corresponding usability constraint is retrieved from the database. (2) The watermark number is determined based on existing requests of that fragment stored in the request log. (3) The pseudo random number generator (PRNG) is seeded using the fragment's secret key and the watermark number. (4) The selection probabilities of the error sub-ranges are computed based on the usability constraint's number of ranges analogue to Table 1.

The following steps are performed for each measurement of the requested fragment. (1) The error range is computed based on the usability constraint's maximum error as well as the sensor's minimum and maximum value. (2) Assuming that preserving the value structure of measurements is an easy way to improve imperceptibility, the error range is further limited by constraints. These constraints (see Table 2) aim to preserve the value structure of the previously watermarked measurement $v_w(t - 1)$, the measurement to be watermarked $v(t)$ and the next measurement $v(t + 1)$. There are two rare exceptions in which preserving the value structure is not desired, namely, if $v_w(t - 1) = v(t)$ and if $v(t) = v(t + 1)$. (3) The error sub-ranges below and above the original value are computed based on the error range. (4) The PRNG selects an error sub-range and then, within the selected error sub-range, the PRNG selects an error.

Table 2. Constraints to preserve the value structure of measurements

Previous measurement value	Next measurement value
if $v_w(t - 1) < v(t)$ then $v_w(t - 1) \leq v_w(t)$	if $v(t) < v(t + 1)$ then $v_w(t) \leq v(t + 1)$
if $v_w(t - 1) > v(t)$ then $v_w(t - 1) \geq v_w(t)$	if $v(t) > v(t + 1)$ then $v_w(t) \geq v(t + 1)$

Task 2 (Embedding). The generated watermark is embedded by adding the watermark to the requested fragment resulting in the watermarked fragment. More precisely, for each measurement of the requested fragment the corresponding error of the watermark is added.

3.3 Watermark Detection

Whenever a suspicious dataset is found somewhere, watermark detection can be applied to determine if the suspicious dataset originates from the data pool and who may have leaked the data. A suspicious dataset is fragmented into a set of suspicious fragments based on sensor and date. For the sake of simplicity, we assume that the timestamps of suspicious datasets are not modified. For each suspicious fragment, the matching original fragment is identified by using fragment similarity search. If a matching fragment is found, the embedded watermark is extracted, and the matching fragment's already embedded watermarks are re-generated. Because the extracted watermark may be noisy due to attacks, possibly matching watermarks are detected by watermark similarity searches. The combination of all matching watermarks of a suspicious dataset reliably identifies potential data leakers. The approach for watermark detection is visualized in Fig. 3 and discussed in the following.

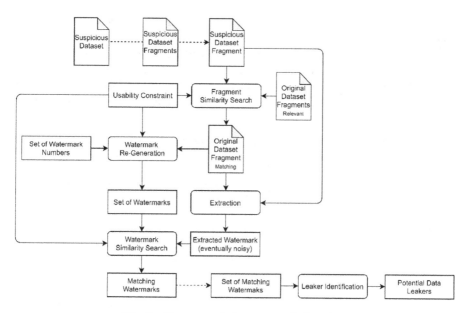

Fig. 3. The approach for watermark detection

Task 1 (Fragment Similarity Search). The fragment similarity search identifies a suspicious fragment's matching original fragment.

The following steps are performed for each suspicious fragment. (1) The suspicious fragment's corresponding usability constraint is retrieved from the database. (2) The set of relevant fragments are retrieved from the database. We used all fragments from the database as relevant fragments. (3) For each relevant fragment, we perform a matching measurements analysis to identify matching measurements between a suspicious and relevant fragment by comparing their measurement timestamps. The result of a matching measurements analysis is a list of suspicious measurements together with their matching original measurement from a relevant fragment. This analysis is necessary because measurements from the suspicious fragment may be removed. (4) For each relevant fragment with the number of matching measurements equal to the suspicious fragment's number of measurements, the fragment similarity is computed using Eq. 1. (5) If the original fragment with the highest similarity exceeds a user-defined threshold, the matching fragment is identified. For the sake of simplicity, we only consider a suspicious fragment's most matching fragment instead of multiple very well matching fragments.

$$Similarity_{Fragment} = \frac{\sum_{i=1}^{matching\ measurements} 1 - \frac{|suspicious\ value_i - original\ value_i|}{sensor's\ maximum\ value}}{number\ of\ suspicious\ measurements} \tag{1}$$

Task 2 (Extraction). The embedded watermark is extracted by subtracting the matching fragment from the suspicious fragment. More precisely, for each measurement of the suspicious fragment the corresponding measurement value of the matching fragment is subtracted.

Task 3 (Watermark Re-Generation). The matching fragment's already embedded watermarks are re-generated based on previous requests. Therefore, the watermark numbers of the matching fragment are retrieved from the request log. The watermarks are re-generated using the watermark generation algorithm from watermark embedding. Additionally, any watermark combinations can be made, e.g. we compute mean watermarks based on a user-defined parameter ("number of colluders") to combine an arbitrary number of watermarks.

Task 4 (Watermark Similarity Search). The watermark similarity search identifies matching watermarks of the extracted watermark. Therefore, the watermark similarity between the extracted watermark and each original watermark is computed using Eq. 2. If the similarity of an original watermark is above a user-defined threshold, it is considered a matching watermark.

$$Similarity_{Watermark} = \frac{\sum_{i=1}^{matching\ measurements} 1 - \frac{|extracted\ error_i - original\ error_i|}{2*maximum\ error}}{number\ of\ matching\ measurements} \quad (2)$$

Task 5 (Leaker Identification). The leaker identification identifies potential data leakers using the matching watermarks from the watermark similarity searches. The dataset leakage probability of a data user is its average watermark similarity. A single fragment may not reliably identify a leaking data user, but multiple fragments do by the watermark combination.

4   Evaluation

In this section we evaluate the watermarking approach regarding configurability, robustness and performance using a proof-of-concept prototype. The proof-of-concept prototype is available online[1].

4.1   Configurability

The usability constraint configures the watermark generation by stating the maximum error, the sensor's minimum and maximum value as well as the number of ranges. The configurability of the watermarking approach ensures that data can be watermarked in a way such that watermarked data remain useful for diagnostic purposes and by any means do not lead to misdiagnosis of patients. In addition, if watermark embedding is configurable based on type and unit of measurement, it could be applied to any kind of sensor data.

Table 3 provides example deviations from the standardized metrics for continuous blood glucose monitoring by Battelino et al. [2] using a looser and a tighter usability constraint. In contrast, using the looser usability constraint results in little deviations in the mean glucose management indicator and glycemic variability, while using the tighter usability constraint results in no deviations in these metrics. The metrics

---

[1] https://github.com/jku-win-dke/iwdw20-prototype.

indicating the time within a range are computed by counting the number of measurements within that range because each measurement is approximately 5 min long. In both cases, using the looser and tighter usability constraint result in high relative deviations. Nevertheless, these deviations may be negligible, considering absolute deviations. Additionally, further tightening the usability constraint may also reduce these deviations.

In summary, the errors introduced by watermarking are configurable by the usability constraint. The tighter the usability constraint, the more imperceptible but less secure the watermark. The determination of an imperceptible but secure usability constraint is one of the major challenges using this approach.

Table 3. Example deviations from standardized metrics using different usability constraints

Metrics	Original	Maximum error = 1.0, Number of ranges = 2		Maximum error = 0.1, Number of ranges = 20	
		Watermarked	Δ	Watermarked	Δ
Mean Glucose	10.2609	10.2762	0.15%	10.2607	0.00%
Glucose Management Indicator	60.9965	61.0685	0.12%	60.9954	0.00%
Glycemic Variability 1	5.4146	5.4109	−0.07%	5.4146	0.00%
Glycemic Variability 2	15.1072	15.1415	0.23%	15.1067	0.00%
Time above Range: > 13.9	54	57	5.56%	54	0.00%
Time above Range: 10.1–13.9	78	74	−5.13%	77	−1.28%
Time in Range: 3.9–10.0	141	142	0.71%	140	−0.71%
Time below Range: 3.0–3.8	6	8	33.33%	7	16.67%
Time below Range: < 3.0	3	3	0.00%	2	−33.33%

4.2 Robustness

The robustness is evaluated regarding rounding value attacks, random value attacks, deletion attacks, subset selection attacks and mean collusion attacks. To this end, we watermarked a single fragment using a usability constraint with a maximum error of 0.5 and a number of ranges of 10. Then, we attacked the watermarked fragment using the attacks listed above with different configurations and let the watermark detection determine fragment similarity and watermark similarity.

The results of these simulations are shown in Tables 4, 5, 6, 7 and 8 with one table per attack method and one line per configuration. The first column shows the configuration parameter, the second column shows the matching fragment similarity (between the suspicious fragment and its matching original fragment) and Columns 3–7 shows the matching watermark similarities for five different watermarks for this fragment (Column Headings 1–5 indicate the watermark numbers). In Simulation 1–4 the to-be matched embedded watermark is that with Watermark 1 (similarities are shown in bold).

The watermarking approach is robust if the matching watermark similarity of the embedded watermark that is to be matched is markedly higher than the similarity with other watermarks.

Simulation 1 (Rounding Value Attacks). The rounding value attack rounds the values of the measurements based on a decimal digit, eventually distorting the watermark. Table 4 provides an exemplary robustness report against rounding value attacks. The watermark similarity decreases notably when rounding the measurements to whole numbers, but this may also exceed the usability constraint. Considering a dataset of multiple fragments, the watermark combination enables reliable identification of the data user.

Simulation 2 (Random Value Attacks). The random value attack changes the values of the measurements randomly based on a maximum error, eventually distorting the watermark. Table 5 provides an exemplary robustness report against random value attacks. The watermark similarity linearly decreases if the introduced maximum error increases. Considering a dataset of multiple fragments, the watermark combination enables reliable identification of the data user.

Simulation 3 (Deletion Attacks). The deletion attack removes measurements based on a certain frequency. Table 6 provides an exemplary robustness report against deletion attacks. Because of the time dependent watermark detection, all matching measurements can be identified, and the watermark similarities are not compromised. The matching fragment similarities differ among each other due to removed measurements.

Table 4. Example robustness against rounding value attacks

Decimal digit	Matching fragment similarity	Matching watermark similarity				
		1	2	3	4	5
4	99.87%	**100.00%**	88.65%	88.10%	89.06%	88.86%
3	99.87%	**99.98%**	88.65%	88.09%	89.06%	88.86%
2	99.87%	**99.77%**	88.65%	88.08%	89.04%	88.88%
1	99.86%	**97.53%**	88.22%	87.62%	88.80%	88.71%
0	99.53%	**76.06%**	72.61%	70.94%	72.35%	72.02%

Table 5. Example robustness against random value attacks

Maximum Error	Matching fragment similarity	Matching watermark similarity				
		1	2	3	4	5
0.1	99.83%	**94.92%**	87.34%	86.85%	87.36%	87.51%
0.2	99.76%	**89.83%**	84.51%	83.78%	84.24%	84.47%
0.3	99.68%	**84.75%**	80.87%	80.00%	80.33%	80.51%
0.4	99.59%	**79.66%**	76.67%	75.88%	75.98%	76.23%
0.5	99.50%	**74.58%**	72.16%	71.40%	71.29%	71.60%

Table 6. Example robustness against deletion attacks

Frequency	Matching fragment similarity	Matching watermark similarity				
		1	2	3	4	5
5	99.86%	**100.00%**	88.39%	87.64%	88.47%	89.11%
4	99.87%	**100.00%**	88.86%	87.74%	89.22%	88.65%
3	99.86%	**100.00%**	88.73%	88.17%	88.96%	88.85%
2	99.86%	**100.00%**	88.01%	88.22%	89.20%	88.79%

Simulation 4 (Subset Selection Attacks). The subset selection attack selects subsets of measurements from each fragment of the dataset based on a certain start and end index. Table 7 provides an exemplary robustness report against subset selection attacks. The watermark similarities are not compromised because of the time dependent watermark detection, while the fragment similarities differ among each other due to missing measurements.

Simulation 5 (Mean Collusion Attacks). The mean collusion attack creates a new dataset based on the mean values of differently watermarked measurements. The watermark detection of a mean colluded dataset is always 100% successful if the correct number of colluders is given to the algorithm due to being able to arbitrary combine watermarks.

We are, however, more interested in providing the watermark similarities of colluded datasets without having to combine watermarks in the watermark detection. Table 8 summarizes the results of the simulation where a different number of colluders work together. In the first configuration two colluders attacked the watermark by computing the mean of their watermarked fragments (with Watermark 1 and 2); in the second configuration three colluders attacked the watermark by computing the mean of their watermarked fragments (with Watermark 1, 2 and 3); and so forth. One can see in the results that even without computing watermark combinations the watermark similarity of each colluder is slightly increased compared to watermark similarities of innocent data users.

Table 7. Example robustness against subset selection attacks

Index	Matching fragment similarity	Matching watermark similarity				
		1	2	3	4	5
100–159	99.87%	**100.00%**	89.97%	89.26%	89.35%	88.00%
100–147	99.85%	**100.00%**	89.43%	87.84%	88.39%	87.07%
100–135	99.84%	**100.00%**	88.29%	85.68%	87.62%	86.51%
100–123	99.81%	**100.00%**	85.88%	84.41%	89.09%	85.35%
100–111	99.79%	**100.00%**	82.87%	88.09%	86.96%	85.86%

Table 8. Example robustness against mean collusion attacks without watermark combinations

Colluders	Matching fragment similarity	Matching watermark similarity				
		1	2	3	4	5
1, 2	99.90%	**94.32%**	**94.32%**	89.21%	90.89%	90.27%
1, 2, 3	99.91%	**93.55%**	**92.92%**	**92.81%**	90.80%	90.82%
1, 2, 3, 4	99.92%	**93.11%**	**92.86%**	**92.23%**	**93.10%**	90.85%
1, 2, 3, 4, 5	99.93%	**93.13%**	**92.62%**	**92.00%**	**92.63%**	**92.68%**

In summary, the approach is robust against rounding and random value attacks at least by the watermark combination of multiple fragments if the attacks do not modify the data exceeding the usability constraint. Because of the time dependent detection, the approach is robust against deletion and subset selection attacks. Nevertheless, the time dependent detection also results in a vulnerability against attacks targeting the time field, e.g. time shifting attacks. Finally, the approach is robust against mean collusion attacks especially if watermark combinations are computed. Thus, in summary, the approach is robust against the evaluated attacking methods.

It should be noted that also other relevant attacks against watermarking of sensor data exist. Imagine, as an example, an attacker who captures data sent by sensors and compares it with the watermarked data obtained from the data platform. In this paper we, however, do not consider such attacks.

4.3   Performance

The performance is evaluated by measuring the time required for watermark embedding and watermark detection. The performance studies were conducted on a Lenovo Thinkpad T470p with a locally installed PostgreSQL[2] database and the file system for data exchange. It should be noted that the time totals in the following two tables are more than the sum of the components because they include computational overhead.

Simulation 6 (Watermark Embedding). The watermark embedding performance depends on the number of requested fragments. Table 9 provides an exemplary performance report of watermark embedding of requested datasets with different sizes. The report shows that the time required for watermark embedding increases almost linearly for every additional requested fragment. Nevertheless, the time required for watermark embedding may be acceptable for practical use.

Table 9. Example performance of watermark embedding

Number of fragments	Database retrieval	Watermark embedding	Dataset providing	Total
1	205 ms	223 ms	109 ms	546 ms
7	329 ms	1 306 ms	573 ms	2 218 ms
29	410 ms	5 611 ms	2 336 ms	8 367 ms

---

[2] https://www.postgresql.org/.

Simulation 7 (Watermark Detection). The watermark detection performance depends on the suspicious dataset's number of fragments, the number of fragments used in the fragment similarity searches, the number of already embedded watermarks of the matching fragments and the number of colluders given to the detection. For the sake of simplicity, we only evaluate the detection performance using a single watermarked fragment. Furthermore, we only evaluate the detection performance without combining watermarks because computing all possible combinations of multiple watermarks quickly escalates. Table 10 provides an exemplary performance report of watermark detection using different settings. Therefore, we randomly generated fragments stored in the database with equal timestamps such that fragment similarity search has to be performed for each generated fragment. We also set up a different number of watermarks for the fragment used for watermark detection such that the watermark similarity search has to be performed for each watermark. The report shows that for every additional fragment and every additional watermark, the detection performance decreases. Nevertheless, the detection performance may be acceptable for practical use.

Table 10. Example performance of watermark detection

Number of fragments	Number of watermarks	Fragmentation	Watermark detection	Total
1	1	50 ms	240 ms	321 ms
1	100	42 ms	559 ms	606 ms
100	1	47 ms	1 663 ms	1 716 ms
100	100	43 ms	1 987 ms	2 035 ms
1 000	1	46 ms	24 690 ms	24 741 ms
1 000	100	40 ms	25 025 ms	25 071 ms

In summary, the performance of watermark embedding and detection is sufficient for practical usage. Considering a physician who typically requests data of a day, week or month, the additional effort for watermark embedding is manageable.

5 Application Scenario: OwnYourData Semantic Containers – My Personal Connected Health

This section describes an implementation and a concrete use case of digital watermarking in *semantic containers* – a framework to exchange data between multiple parties, which is promoted by OwnYourData as a privacy-preserving way of sharing data. OwnYourData is an Austrian non-profit organization aiming to foster sharing of personal data. Recently, OwnYourData was awarded the status of a MyData Operator[3] in recognition of providing a human-centric infrastructure for personal data management and sharing.

---

[3] https://mydata.org/operators/.

Building on the semantic container approach originally developed for decentralized aeronautical information management [3], OwnYourData has further developed Semantic Containers into a data mobility platform to exchange data in a secure and traceable manner. An OwnYourData Semantic Container is a package containing a dataset, semantic metadata describing the dataset and processing capabilities, together with all the software necessary to interact with the data.

With 425 million adult people diagnosed with Diabetes worldwide this growing epidemic requires adequate resources – including a data-driven approach for managing the complex drug adjustment to control blood glucose levels. Self-monitoring devices of blood glucose provide comprehensive insight for physicians, researchers, and the pharmaceutical industry but at the same time require measures to protect this personal identifiable information when sharing with individuals and companies. In the course of the EU-funded project MyPCH[4] this challenge was addressed using Semantic Containers as data sharing platform.

The Data Donation Dataflow[5] developed in the course of the MyPCH project demonstrates extracting periodic blood glucose measurement from a person with diabetes through Tidepool[6] and storing this data in a local Semantic Container. Any data request to this container requires authentication and provides a unique fingerprint (i.e., watermark) applied to the response. In case the receiving party leaks the received dataset and an unauthorized dataset appears on the internet the person with diabetes can match this dataset against the data in the Semantic Container to identify similarity with the original data and then check if individual fragments (data from a specified day) match the applied digital watermarking for all authenticated accounts.

To detect unauthorized distribution of Semantic Containers or of the data they contain, the watermarking approach presented in this paper was implemented as native capability within the Semantic Container framework, i.e., being part of the Semantic Container base image[7]. The Semantic Container core functionality including digital watermarking is provided as a REST API[8]. The ability to track unauthorized distribution of data is the basis for trust in the platform and should encourage users to share data.

6   Related Work

In this section we give a brief overview of the state of the art and how our settings and design options relate to it. With the health sensors producing time-indexed sequences of measurements, personal health sensor data may be regarded as time series data, c.f. [4]. Typically, watermarking is used for data leakage detection [5].

---

[4] https://wiki.geant.org/display/NGITrust/Funded+Projects+Call+1.

[5] https://github.com/sem-con/sc-diabetes/tree/master/dataflows/Data_Donation.

[6] https://www.tidepool.org/.

[7] https://github.com/sem-con/sc-base.

[8] https://api-docs.ownyourdata.eu/semcon/.

Existing watermarking techniques for time series data focus on, among others, tamper proofing and authentication, as shown in a review by Panah et al. [4]. We use, however, watermarking for avoiding unauthorized re-sharing of personal health sensor data. Furthermore, blind watermarking techniques of biomedical time series data has been proposed by Duy et al. [6] and Pham et al. [7], respectively. Considering the assumed setting with personal health sensor data stored in a database not being modified, we use informed watermark embedding to improve the imperceptibility of the watermark and we use non-blind watermark detection to improve detection performance [1]. Ayday et al. [8] proposed a collusion-secure watermarking technique of sequential data (including time series data) for data leakage detection. In contrast to Ayday et al. [8], we aim to make every single measurement that is shared part of the watermark to increase robustness against certain attacks but still considering preserving the usability of the data.

7 Summary and Future Work

In this paper, we proposed and evaluated a watermarking approach for data leakage detection of personal health sensor data. This approach may also be applicable to different kinds of sensor data because it is based on watermarking only the value fields of measurements. The approach is informed, meaning that both watermark embedding and detection take advantage of an existing copy of the original data. In addition, the watermark embedding is configurable by a usability constraint which depends on the sensor type of the measurements being watermarked. The approach is also robust against several attacks including mean collusion attacks. The performance of watermark embedding and detection is sufficient for practical use.

The watermarking approach provides high extensibility and adaptability because the algorithms for watermark embedding and detection can be arbitrarily extended or adapted. In addition, the watermark detection can be implemented or further improved even if the watermarking system is already in production.

Future work may improve the algorithms of watermark embedding and detection. In case of watermark embedding, especially watermark generation may be improved to enable fast generation of maximally different and very well matching watermarks. In case of watermark detection, improvement can be made by more time-independent similarity searches, criteria to limit the number of relevant fragments for fragment similarity search and considering multiple matching fragments.

Acknowledgments. Part of this work was conducted as part of the MyPCH project. This project received funding from the EU's Horizon 2020 program for research and innovation, NGI_Trust funds via the Grant Agreement Number 825618.

References

1. Cox, I., Miller, M., Bloom, J., Fridrich, J., Kalker, T.: Digital Watermarking and Steganography, 2nd edn. Morgan Kaufmann Publishers Inc., San Francisco (2007)
2. Battelino, T., et al.: Clinical targets for continuous glucose monitoring data interpretation: recommendations from the international consensus on time in range. Diab. Care **42**(8), 1593–1603 (2019)
3. Gringinger, E., Schuetz, C., Neumayr, B., Schrefl, M., Wilson, S.: Towards a value-added information layer for SWIM: the semantic container approach. In: 2018 Integrated Communications, Navigation, Surveillance Conference (ICNS), Herndon, VA, pp. 1–20 (2018)
4. Panah, A.S., Van Schyndel, R., Sellis, T., Bertino, E.: On the properties of non-media digital watermarking: a review of state of the art techniques. IEEE Access **4**, 2670–2704 (2016)
5. Papadimitriou, P., Garcia-Molina, H.: Data leakage detection. IEEE Trans. Knowl. Data Eng. **23**(1), 51–63 (2011)
6. Duy, T.P., Tran, D., Ma, W.: An intelligent learning-based watermarking scheme for outsourced biomedical time series data. In: 2017 International Joint Conference on Neural Networks (IJCNN), Anchorage, pp. 4408–4415 (2017)
7. Pham, T.D., Tran, D., Ma, W.: Ownership protection of outsourced biomedical time series data based on optimal watermarking scheme in data mining. Australas. J. Inf. Syst. **21** (2017). https://doi.org/10.3127/ajis.v21i0.1541
8. Ayday, E., Yilmaz, E., Yilmaz, A.: Robust optimization-based watermarking scheme for sequential data. In: 22nd International Symposium on Research in Attacks, Intrusions and Defenses, pp. 323–336 (2019)

Complete Quality Preserving Data Hiding in Animated GIF with Reversibility and Scalable Capacity Functionalities

KokSheik Wong[1]($\boxtimes$), Mohamed N. M. Nazeeb[1], and Jean-Luc Dugelay[2]

[1] Monash University Malaysia, Subang Jaya, Selangor, Malaysia
wong.koksheik@monash.edu, mnmoh48@student.monash.edu
[2] EURECOM, Sophia Antipolis, France
jean-luc.dugelay@eurecom.fr

Abstract. A technique is put forward to hide data into an animated GIF by exploiting the transparent pixels. Specifically, a new frame is crafted based on the data to be embedded. The newly crafted frame is inserted between 2 existing frames, and the delay time of the affected frames are adjusted accordingly to achieve complete imperceptibility. To the best of our knowledge, this is the first attempt to hide data into an animated GIF by exploiting the transparent pixel. Irregardless of the characteristics of the animated GIF image, the proposed method can completely preserve the quality of the image before and after hiding data. The hiding capacity achieved by the proposed method is scalable, where more information can be embedded by introducing more frames into the animated GIF. While file size expansion is inevitable, reverse zero run length is adopted to suppress the expansion. The proposed method is reversible, i.e., the original image can be recovered.

Keywords: Transparent pixel · Animated GIF · Complete quality preservation · Data hiding · Reversible

1 Introduction

Graphic interface format (GIF) is a highly portable and platform-independent image file format designed to show moving pictures through low bandwidth Internet. It was developed by CompuServe in 1987, where further innovations such as *dirty rectangular* and *transparent pixel* took place after the disclosure of the GIF 89a specifications [1]. Although animated GIF contains no sound/voice, the short visual content shows dynamic content, tells story, and captures emotion [4].

The popularity of animated GIF has been decaying, but recently social networking service platforms and online advertisers are making good use of animated GIFs despite broadband network connectivity. These creative utilizations of animated GIF give new life to the originally dull image, including the transition of different combinations of outfit/shoes on the same model, handbag of a specific model in different colors, to name a few. Furthermore, animated GIFs

© Springer Nature Switzerland AG 2021
X. Zhao et al. (Eds.): IWDW 2020, LNCS 12617, pp. 125–135, 2021.
https://doi.org/10.1007/978-3-030-69449-4_10

can be easily generated nowadays thanks for the availability of freely available encoder in many platforms, including online websites. There are also dedicated websites to blog about, share, search, and create animated GIFs [2,5]. Moreover, users also use animated GIFs in instant messaging platform and online forum to show reactions or emotions.

Due to its popularity and large number in existence, many data hiding methods are designed to better manage GIFs over the years. Traditionally, data is hidden into a digital content such as image to convey secret message [6,10]. One of the earliest techniques designed for GIF is proposed by Kwan, where the color palette is arranged in certain way to convey a secret message [8]. However, the hiding capacity is low. In another technique called EzStego, Machado [9] proposed to analyze the color palette of a GIF image and sort the indices based on luminance. If an index needs to be replaced for hiding data, the nearby indices (post- sorting) are considered. Later, Fridrich et al. [3] proposed to match the parity of the sum of RGB triplet values to the data bit. The nearest RGB triplet with matching sum is selected to represent the message bit. Data can also be hidden without causing any distortion (i.e., complete quality preservation [16]), but the requirement is to start with a GIF with at least 1 un-referenced indice. Kim et al. is able to hide up to 8 bits per pixel without causing distortion when there at least 128 un-referenced indices [7]. Recently, Wang et al. put forward a technique to quantize colors in GIF [14]. Two similar colors C1 and C2 in the color palette are combined by taking their weighted average to generate a new color, where the notion of similarity is defined by some *risk* function. Pixels having the index value of C1 or C2 are manipulated to hide data.

Although there are techniques designed to hide data into animated GIF, they are treating each frame as a static image, where existing techniques such as EzStego [9] and Fridrich et al.'s method [3] are deployed to hide data into the selected frames. In other words, the conventional techniques either modify the pixel index, color table entries, or the combination of both, where distortion is inevitable. In spite the fact that LZW compression is exploited to hide data in GIF [12], other parts of the GIF structure remain unexplored, particularly the parts related to *animation* in GIF. Therefore, in this work, we propose to hide data into an animated GIF file, where new frames are crafted based on the data to be hidden. To the best of our knowledge, our technique is the first of its kind to hide data by inserting new frames and using transparent pixel.

While the conventional techniques surveyed above are mostly designed for steganography, our proposed method can be utilized in the applications of data hiding such as fragile watermark for tamper detection and annotation. In addition, one may envisage a spectacular demo by using the proposed method in animated GIF thanks to its scalable capacity functionality. For example, a binary animation can be hidden in an animated GIF, which is apparently normal.

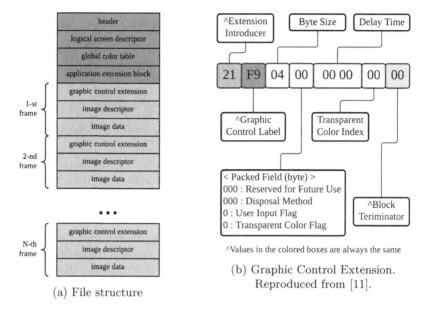

(a) File structure

(b) Graphic Control Extension.
Reproduced from [11].

Fig. 1. File structure of animated GIF and its graphic control extension.

2 Overview of GIF File Structure

Figure 1(a) shows the structure of a GIF file, which consists of protocol blocks (for set-ups) and sub-blocks of graphics. Specifically, an animated GIF A of dimension $M \times N$ pixels consists of an assemble of frames A_f so that $A = \{A_f\}$ for $f = 1, 2, \cdots, F$, where F is the total number of frames. The logical screen descriptor contains information such logical screen width and height, background color index, etc. [1]. On the other hand, the global color table consists of 256 entries of RGB-triplets, and there is a function C that maps index to integer RBG-triplet, i.e., $C : [0, 255] \rightarrow [0, 255] \times [0, 255] \times [0, 255]$. To facilitate the presentation, let $A_f(x, y) \in \{0, 1, \cdots, 255\}$ denote the index at position (x, y) within frame A_f, where $1 \le x \le M$ and $1 \le y \le N$. Each frame A_f consists of three data blocks, namely: graphic control extension, image descriptor, and image data - see Fig. 1(b) [11].

Next, we focus on *Transparent Color Flag* (TCF) within the *Packed Field* and *Transparent Color Index* (TCI). When TCF is set to TRUE, it enables an index to be utilized as the transparent pixel, where color from a previous frame is rendered instead of the color associated with the index. When TCI = τ_f =← 169 for example, the index '169' is reserved and utilized for transparent pixel. Therefore, if $A_f(x, y) = 169 = A9_{16}$, the color $C(169)$ (i.e., triplet of RBG value) will not be displayed at position (x, y) in frame A_f. Instead, the color from the same location in the previous frame, i.e., $A_{f-1}(x, y)$, will be rendered. The transparent pixel concept is introduced for compression purposes. Although its performance varies depending on the characteristics of the animated GIF, a

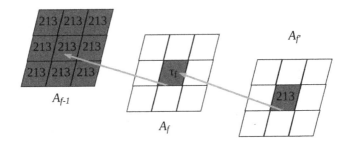

Fig. 2. Illustration of $A_{f'}(x_0, y_0)$ referring to $A_f(x_0, y_0) = \tau_f$, i.e., transparent pixel. The actual color (i.e., index 213) is traced and retrieved from A_{f-1}. Here, (x_0, y_0) refer to the center of the 3×3 image block.

compression ratio of 1.63:1 is reported in [13]. On the other hand, the disposable method informs the decoder what to do with the current frame A_f when the decoder moves on to the next frame A_{f+1}. A value of '0' implies that the image is static and the decoder cannot draw anything on top of it. This value is used for non-animated (i.e., static) GIF. On the other hand, a value of '1' informs the decoder to leave the current image on screen and draw the next image on top of it. There are other modes of operation but we omit the presentation here due to space limitation. The length of display for each frame A_f is controlled by using the value as specified in the *Delay Time* field (denoted by d_f). Basically a frame A_f will stay on the screen for d_f centi-seconds (i.e., 1/100 of a second). For the purpose of this work, we set TCF to TRUE, and use '1' for the disposable method.

3   Proposed Data Hiding Method

In this section, we first propose a pre-processing step to prepare the animated GIF A for data hiding purpose. The actual data hiding and extraction processes are then put forward. Finally, we explain how reverse zerorun length encoding [16] is adopted to overcome the problem of file size increment.

3.1   Pre-processing

A new frame $A_{f'}$ is created and inserted between A_f and A_{f+1} to facilitate data hiding. Each pixel index $A_{f'}(x, y)$ is eventually modified to hide data. Specifically, a new frame $A_{f'}$ is created by copying all indices from A_f, i.e.,

$$A_{f'}(x, y) \leftarrow A_f(x, y). \tag{1}$$

Here, the same indices (hence colors) are copied from A_f to ensure imperceptibility of the newly inserted frame $A_{f'}$. However, $A_{f'}$ requires additional treatment when there is at least one transparent pixel occurring in A_f, or more precisely, $|\{(x, y) : A_f(x, y) = \tau_f\}| > 0$, where $|X|$ refers to the cardinality of the set X.

Specifically, due to the simple duplication process of Eq. (1), an issue arises when $A_f(x_0, y_0) = \tau_f \in [0, 255]$, where the index τ_f is defined as the transparent pixel in frame A_f. In other words, $A_f(x_0, y_0) = \tau_f$ means that an actual color in an earlier frame, i.e., $A_\alpha(x_0, y_0)$, is referred for display, where $\alpha < f$. In the event we set $\tau_{f'} \neq \tau_f$ (i.e., we use different indices to define the transparent pixels in A_f and $A_{f'}$), the actual color of $C(\tau_f)$ will be displayed at $A_{f'}(x_0, y_0)$, instead of the color in an earlier frame, namely, $A_\alpha(x_0, y_0)$.

To overcome the aforementioned issue, the main objective of the pre-processing is to eliminate *all* transparent pixels in the newly created frame $A_{f'}$, where every occurrence of the transparent index τ_f will be substituted by the actual color (i.e., a RGB-triplet referred by an index) in the earlier frame A_α for $\alpha < f$. Figure 2 illustrates a scenario where $A_f(x_0, y_0) = \tau_f$ is a transparent pixel, which refers to the color shown at position $A_{f-1}(x_0, y_0)$. Hence, the actual color $C(213)$ shown at position $A_{f-1}(x_0, y_0)$ is traced and copied, in other words, $A_{f'}(x_0, y_0) = 213$. The process is repeated to eliminate all transparent pixels in $A_{f'}$. Eventually, the newly added frame $A_{f'}$ consists entirely of indices to actual RGB-triplets without any transparent pixels. In other words, $|\{A'_f(x, y) = \tau_f\}| = 0$.

Next, we have 2 scenarios to manage, namely, A_f has a defined transparent pixel, and A_f does not have a defined transparent pixel. For the former scenario, we continue to utilize the same transparent pixel, i.e, $\tau_f \leftarrow \tau_f$, instead of finding another index for such purpose. On the other hand, for the latter situation, we have to choose the transparent pixel $\tau_{f'}$ carefully. Specifically, all indices in frame A_f are scanned and the histogram H_f of the indices is constructed. Let $H_f(i)$ denote the frequency of occurrences for the index i, where $0 \leq i \leq 255$. We select the index i_0 such that $H_f(i_0)$ is the minimum (i.e., occurring the least in A_f), and set $\tau_{f'} \leftarrow i_0$. Note that in practice, a GIF image does not utilize all 256 indices. Therefore, in general, $H_f(i_0) = 0$ holds true and we can hide 1 bit per pixel (bpp). On the other hand, when $H_f(i_0) > 0$, we skip the positions $A_{f'}(x, y)$ (i.e., newly added frame) for data hiding when $A_f(x, y) = i_0$ (i.e., original frame). Here, we loose exactly $H_f(i_0)$ number of pixel locations for data hiding, and the embedding reduces to $1 - H_f(i_0)/M/N$ bpp.

In both cases, data hiding can take place, where defining a new transparent pixel will not confuse the extraction process with the introduction of *usable* and *non-usable* positions in Sect. 3.2.

3.2   Data Hiding

To hide data, the new frame $A_{f'}$ is compared with A_f at each pixel location. Specifically, the position $A_{f'}(x, y)$ is skipped and we call it *non-usable* if the following two conditions are true simultaneously:

$$\tau_f \neq \tau_{f'} \tag{2}$$

$$A_f(x, y) = \tau_{f'}. \tag{3}$$

Note that such a decision is made to avoid ambiguity during data extraction because due to the simple duplication process (i.e., Eq. (1)), we cannot differentiate whether $A_{f'}(x, y) = \tau_{f'}$ is encoding '0' (see Eq. (4)), which is modified from $A_f(x, y)$, or it is actually the original index for that pixel location. Therefore, we skip these positions.

On the other hand, $A_{f'}(x, y)$ is called *usable* and it is exploited to hide data by using the basic rules below:

$$A_{f'}(x, y) \leftarrow \begin{cases} \tau_{f'} & \text{if } m_k = 0; \\ \text{'No change' otherwise.} \end{cases} \tag{4}$$

Here, the payload m is a binary sequence $\{m_k\} \in \{0, 1\}$. The encoding rule basically utilizes the transparent pixel index to encode '0', and utilizes the original index to encode '1'. The process is repeated for each position (x, y) in the frame $A_{f'}$ in the raster scanning order.

In order to maintain the length of the original animated GIF, the delay time for frame A_f and $A_{f'}$ need to be adjusted. Specifically, we set $d_{f'} \leftarrow \lfloor d_f / 2 \rfloor$, where $\lfloor z \rfloor$ refers to the largest integer smaller than or equal to z. Next, we update $d_f \leftarrow d_f - d_{f'}$. Essentially, the proposed method splits A_f into 2 frames, both having the exact same pixel values on screen, and the overall display duration (i.e., delay time) remains unchanged. Since the exact same pixel values are displayed for the same duration, the quality is completely preserved. In other words, the pixel values rendered from the original and processed (embedded with data) animated GIF images are exactly the same, and these pixels appear on the screen for exactly the same duration. In fact, the duration d_f and $d_{f'}$ can be further manipulated to hide data, which will be explored as our future work.

By inserting a new frame between every 2 consecutive original frames, we are increasing the number of frames from F to $2F - 1$. In fact, to improve hiding capacity, more new frames can be generated and inserted between any two consecutive frames, including the pairs A_f and $A_{f'}$ as well as $A_{f'}$ and A_{f+1}. This process can be repeated as long as all delay times (i.e., d_f and $d_{f'}$) remain ≥ 0.02s, which is the smallest permissible value allowed by web browser.

3.3 Data Extraction and Reversibility

To extract data from the animated GIF embedded with data A', the inserted frames $A'_{f'}$ are first identified. This process can be achieved by some prearrangement, for example, a new frame is always added between 2 original frames (i.e., A_f and A_{f+1}), and hence the odd numbered frames in A' are the newly inserted frames. The status (being *usable* or *unusable*) of each position in $A'_{f'}$ is verified by referring to Eq. (2) and (3). The sequence of embedded bits in $A'_{f'}$ is extracted from the *usable* locations by producing '1' when $A'_{f'}(x, y) = A'_f(x, y)$ or '0' when $A'_{f'}(x, y) = \tau_{f'}$. The process is repeated for all inserted frames $A'_{f'}$.

The proposed method is obviously reversible, where the newly added frames $A_{f'}$ can be removed and the original delay time d_f can be reassigned to recover the original animated GIF image.

3.4   Reducing File Size Increment

When a new frame is inserted, file size is inevitably increased. To reduce file size expansion, the *reverse zerorun length* (RZL) encoding technique [16] is adopted. Note that for each newly created frame $A_{f'}$, prior to any modifications due to data hiding purposes, $A_{f'}$ encodes a sequence of 1's with length $M \times N$ (or slightly lesser depending on $H_f(\tau_{f'})$ in A_f). Instead of using Eq. (4) to hide data directly, the message is first pre-processed. Specifically, for each newly created frame $A_{f'}$, the data to be hidden ϕ_f is divided into segments each of length k-bits, i.e., $\phi_f = [\phi_f^1, \phi_f^2, \cdots, \phi_f^D]$ where $D = |\phi_f|/k$. Here, each segment ϕ_f^k is of length k bits except for ϕ_f^D, which can assume a length $\leq k$ bits.

Next, the decimal equivalent of ϕ_f^i, denoted by d_f^i, is computed and hence $0 \leq d_f^i \leq 2^k - 1$. Subsequently, d_f^i is utilized to generate a new segment μ_f^i for $i = 1, 2, \cdots, D$. The segments μ_f^i are generated as follows:

$$\mu_f^i = \underbrace{00\cdots0}_{d_f^i}1, \tag{5}$$

which is sequence of d_f^i zeros, followed by a '1' that serves as a delimiter. Note that μ_f^i is of variable length. The new representation of the message, i.e., μ_f^i, is then embedded by using Eq. (4). If the newly inserted frame $A_{f'}$ is unable to hide all segments μ_f^i, a new frame $A_{f''}$ can be inserted between $A_{f'}$ and A_{f+1} to create more room for data hiding.

To extract the hidden data segment encoded in the RZL format, the sequence of 0's and 1's are first extracted from all *usable* pixel locations, where $A_{f'}'(x, y) = \tau_{f'}$ outputs a '0', otherwise a '1'. The extracted sequence is then analyzed, where the number of 0's preceeding the value 1 is counted and converted into a binary number with k-bits. For example, the following sequence of 19 bits are extracted from 19 *usable* pixel locations:

$$\underbrace{00000\,1}_{5}\underbrace{00000000\,1}_{8}\underbrace{000\,1}_{3}. \tag{6}$$

The corresponding decimal values 5, 8 and 3, are converted into binary numbers 101, 1000 and 11, respectively. Finally, leading zeros are injected to make up the number of bits (i.e., length) for each segment. Suppose $k = 6$, then the segments become 000101, 001000, and 000011, respectively.

4   Experiments

The proposed data hiding method is implemented in Python. 8 animated GIFs from the world wide web are considered for experiment purpose, where the first frame of each animated GIF is shown in Fig. 3. These GIFs are either generated by using graphic software or merging frames/scenes from video recording.

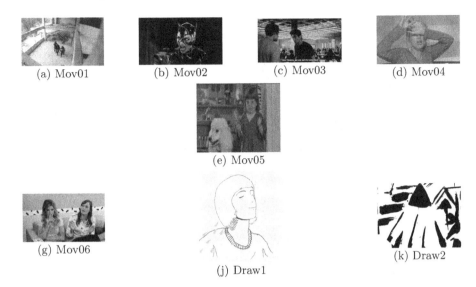

(a) Mov01   (b) Mov02   (c) Mov03   (d) Mov04

(e) Mov05

(g) Mov06   (k) Draw2

(j) Draw1

Fig. 3. First frame of each animated GIF considered for experiment.

Additional information of these animated GIFs can be found in Table 1. They are also made available online at [15] for reproducibility and future comparison purpose. Google Chrome (version 73.0.3683.103), Safari (version 12.0.2), Firefox (version 66.0.3) and Photos (system viewer for Windows 10) are utilized to display the animated GIFs. The processed animated GIFs can be viewed by using the aforementioned browsers, and this observation also confirms that the processed images are format compliant. It is verified that the hidden data can be extracted by checking the status (i.e., *usable* or *non-usable*) of each pixel locations using Eq. (3). By visual inspection, the GIFs appear to be identical before and after hiding data. Unless specify otherwise, $F - 1$ new frames are introduced to an animated GIF with F frames. Although frames of different sizes can be created, for experiment purposes, the dimension of each new frame A'_f is set to be the same as that of the original frame A_f in the respective GIF.

Note that we do not evaluate image quality by using metrics such as MSE or SSIM because the exact same RGB-triplet is rendered at each position, i.e., complete quality preservation. It is also noteworthy that, irregardless of the statitics of the host image, the proposed method can surely embed data without causing any quality degradation, while the conventional methods degrade the image quality because 2 color indices are combined to free up an index [14] or the pixel value is modified by mapping it to different color index [3].

Table 1. Basic information about the animated GIFs considered for experiments.

GIF filename	Image dimensions	Total frames (original)	Bit stream size (KBytes)							
			Original	Basic	RZL					
					$k = 2$	$k = 3$	$k = 4$	$k = 5$	$k = 6$	
Mov01	350 × 196	71	2,000	3,958	3,303	3,022	2,722	2,463	2,283
Mov02	499 × 273	17	472	1,436	1,111	960	805	682	596
Mov03	500 × 254	15	1,276	2,259	1,934	1,776	1,622	1,492	1,404
Mov04	250 × 141	25	505	958	821	750	673	613	568
Mov05	260 × 208	23	485	1,036	875	792	697	624	569
Mov06	480 × 270	17	1,301	2,404	2,054	1,893	1,708	1,558	1,453
Draw1	400 × 426	6	665	1,087	934	865	798	747	714
Draw2	670 × 503	52	2,488	6,754	5,334	4,894	4,262	3,685	3,239

4.1   Hiding Capacity

The number of bits that can be inserted into each GIF (i.e., payload) of each image is recorded in Table 2. All animated GIFs considered in this work consist of transparent pixel in each frame, therefore we could conveniently set $\tau_{f'} \leftarrow \tau_f$. As a result, all pixel locations are *usable*. When using the basic rules (i.e., Eq. (4)) to hide data, the payload is 1 bpp for each added frame. When using RZL, the payload decreases when the parameter k is increased. The payload decreases by a factor of $\sim$3 when k increases from 2 to 6. Although the payload achieved by RZL(6) is slightly less than 1/5 of *Basic*, the suppression of bit stream expansion is significant. On the other hand, the conventional methods are all limited by the number of frames as well as number of pixels in the animated GIF to hide data, i.e., non-scalable, while the proposed method is scalable at the expense of larger file size.

4.2   File Size Expansion

It is expected that the bit stream size will expand since new frames are introduced to hide data. The results are recorded in Table 1, with and without the implementation of RZL. When using the basic rule to hide data, the average expansion of bit stream size is 66.9%, which is reasonable since the number of frames is almost doubled. For completion of discussion, we also record the embedding efficiency η, which is defined as the number of embedded payload bits for every increased bit in the host image. Specifically, we consider the ratio of $\kappa(A, k)$ to $\Delta(A, A')$, where $\kappa(A, k)$ is the embedding capacity for the image A when using the parameter k, and $\Delta(A, A') = FS(A') - FS(A)$ refers to the file size difference between the original image A and the processed image A'. Here, higher value of η implies better performance, and vice versa. The average result $\bar{\eta}$ is recorded in the last row of Table 2 for $k = 1, 2, \cdots, 6$. On average, the host animated GIF image spends $1/0.28 \sim 3.8$ bits for hiding 1 bit of the payload.

On the other hand, when RZL is adopted, it is obvious that the expansion in bit stream size is suppressed, where the effect is more apparent for larger k.

Table 2. Embedding capacity (KBytes) for various k value after applying reserve zero run length encoding [16]

Image	Basic	$k = 2$	$k = 3$	$k = 4$	$k = 5$	$k = 6$
Mov01	586	334	317	247	165	101
Mov02	266	152	144	112	75	47
Mov03	217	123	117	91	61	38
Mov04	103	59	55	42	29	18
Mov05	145	83	79	61	42	26
Mov06	253	144	138	105	71	44
Draw1	104	59	56	43	29	18
Draw2	2,098	1,198	1,142	885	600	375
$\bar{\eta}$		0.28	0.24	0.29	0.32	0.34	0.35

Specifically, the average bit stream size expansion drops from 44.8% to 16.7% when k increases from 2 to 6. However, as noted in the previous sub-section, payload is reduced when k increases. Interestingly, the embedding efficiency decreases initially when RZL is adopted (i.e., $k = 2$), but the performance improves steadily after for $k > 2$. A potential influence to the performance is the LZW compression process, which is part of the GIF standard. This will also be explored as one of our future work.

In contrast, the conventional mostly maintains the file size, with small variation due data hiding. While the proposed method and conventional methods cited in this paper have their pros and cons, they can be combined to complement one an another. The combined deployment will be further explored as our future work.

5   Conclusions

In this work, transparent pixel in animated GIF is manipulated to hide data. Specifically, a new frame is introduced between 2 original frames, and each pixel location is manipulated to hide data. When a location is assigned the transparent pixel index, color from the previous frame is copied and rendered. Delay time of each frame is adjusted accordingly to ensure the duration of the animated GIF remains unchanged. Complete quality preservation is achieved irregardless of the characteristics of the animated GIF image, and the proposed method is reversible where the original animated GIF can be perfectly restored. Experiments suggest that data can be hidden into and extracted from the animated GIF.

In future work, we want to explore how the delay time parameter in each frame can be utilized to hide data. Furthermore, the joint utilization of the proposed and the conventional data hiding methods will be investigated. The influence of LZW compression in GIF on the file size increment due to data hiding will be also be investigated.

Acknowledgement. This work was supported in part by the Fundamental Research Grant Scheme (FRGS) MoHE Grant under project - Recovery of missing coefficients - fundamentals to applications (FRGS/1/2018/ICT02/MUSM/02/2) and in part by EU Horizon 2020 - Marie Sklodowska-Curie Action through the project entitled Computer Vision Enabled Multimedia Forensics and People Identification (Project No. 690907, Acronym: IDENTITY).

References

1. Graphics Interchange Format: Version 89a (1990). https://www.w3.org/Graphics/GIF/spec-gif89a.txt. Accessed 3 Mar 2019
2. Cooke, J., Chung, A.: Giphy. https://giphy.com/. Accessed 3 Mar 2019
3. Fridrich, J.: A new steganographic method for palette-based images. In: PICS 1999: Proceedings of the Conference on Image Processing, Image Quality and Image Capture Systems (PICS-99), Savannah, Georgia, USA, 25–28 April 1999, pp. 285–289. IS&T - The Society for Imaging Science and Technology (1999)
4. Gygli, M., Soleymani, M.: Analyzing and predicting GIF interestingness. In: Proceedings of the 24th ACM International Conference on Multimedia, MM 2016, pp. 122–126. ACM, New York (2016). https://doi.org/10.1145/2964284.2967195. http://doi.acm.org.ezproxy.lib.monash.edu.au/10.1145/2964284.2967195
5. Karp, D.: Tumblr. https://www.tumblr.com/. Accessed 3 Mar 2019
6. Katzenbeisser, S., Petitcolas, F.A. (eds.): Information Hiding Techniques for Steganography and Digital Watermarking, 1st edn. Artech House Inc., Norwood (2000)
7. Kim, S., Cheng, Z., Yoo, K.: A new steganography scheme based on an index-color image. In: 2009 Sixth International Conference on Information Technology: New Generations, pp. 376–381, April 2009. https://doi.org/10.1109/ITNG.2009.119
8. Kwan, M.: GIF colormap steganography (1998). http://www.darkside.com.au/gifshuffle/
9. Machado, R.: Ezstego (1997). http://www.stego.com/
10. Pan, Z., Wang, L.: Novel reversible data hiding scheme for two-stage vqcompressed images based on search-order coding. J. Vis. Commun. Image Represent. **50**, 186–198 (2018). https://doi.org/10.1016/j.jvcir.2017.11.020. http://www.sciencedirect.com/science/article/pii/S1047320317302286
11. Raymond, E.S.: What's in a GIF - animation and transparency (2012). http://giflib.sourceforge.net/whatsinagif/animation_and_transparency.html
12. Shim, H.J., Jeon, B.: DH-LZW: lossless data hiding in LZW compression. In: 2004 International Conference on Image Processing, ICIP 2004, vol. 4, pp. 2195–2198, October 2004. https://doi.org/10.1109/ICIP.2004.1421532
13. Thyssen, A.: ImageMagick v6 examples - animation basics (2004). https://imagemagick.org/Usage/anim_basics/
14. Wang, X., Yao, T., Li, C.T.: A palette-based image steganographic method using colour quantisation. In: IEEE International Conference on Image Processing 2005, vol. 2, p. II-1090, September 2005. https://doi.org/10.1109/ICIP.2005.1530249
15. Wong, K., Nazeeb, M.N.M., Dugelay, J.L.: Test animated GIFs (2020). http://bit.ly/2IEx26N
16. Wong, K., Tanaka, K., Takagi, K., Nakajima, Y.: Complete video quality-preserving data hiding. IEEE Trans. Circ. Syst. Video Technol. **19**(10), 1499–1512 (2009). https://doi.org/10.1109/TCSVT.2009.2022781

Visible Reversible Watermarking for 3D Models Based on Mesh Subdivision

Fei Peng[1]([✉]), Wenjie Qian[1], and Min Long[2]

[1] College of Computer Science and Electronic Engineering, Hunan University,
Changsha 410082, Hunan, China
eepengf@gmail.com
[2] College of Computer and Communication Engineering,
Changsha University of Science and Technology, Changsha 410114, Hunan, China

Abstract. Visible reversible watermark can clearly identify the copyright of digital works, and can completely remove the watermark when it is needed. It can be utilized for visibly labeling the copyright of digital host. To protect the copyright of 3D mesh model, a visible reversible watermarking based on mesh subdivision is proposed in this paper. First, the smooth area of the 3D mesh model is projected onto the 2D plane for cropping and subdivision to achieve the embedding of the visible watermark. Then, by comparing the vertex information of the original model and the model with visible watermark, the index of the visible watermark vertex is obtained, and it is embedded as a reversible watermark to the rest of the model. The visible watermark can be removed by deleting the vertices corresponding to the reversible watermark. Experimental results and analysis show that the visible watermark and the host are tightly integrated, and it is robust to translation, rotation, scaling, mesh subdivision, and mesh smoothing. When unauthorized users try to delete the watermark or steal the 3D mesh model, the host will be destroyed and the model will be unusable. It has potential application in the copyright protection of the 3D mesh model.

Keywords: 3D mesh model · Visible reversible watermark · Mesh subdivision · Copyright protection

1 Introduction

In recent years, with the rapid development of computer-aided design, virtual reality, 3D model processing and other technologies, 3D mesh models have played an important role in industrial manufacturing, architectural design, cultural relics protection, animation and game model design. For 3D mesh model, it often faces three threats: data privacy leakage, malicious tampering and copyright disputes. If it is not well protected, it will result in risks of data security.

This work was supported in part by projects supported by National Natural Science Foundation of China (Grant No. 92067104, U1936115, 62072055).

X. Zhao et al. (Eds.): IWDW 2020, LNCS 12617, pp. 136–149, 2021.
https://doi.org/10.1007/978-3-030-69449-4_11

As an important method of copyright protection for digital products, digital watermarking has been extensively developed and applied. Compared with invisible watermark, visible watermark allows the observer to perceive the embedded watermark information. It is suitable for identifying the copyright of 3D model, and preventing illegal use of protected data, etc. For example, embedding company logo or some useful information on the 3D model can clearly identify the model's belonging. Moreover, visible reversible watermarking can visually protect the copyright of the host, and restore the original host by removing the watermark without loss. They generally use the redundancy of the host to compress and make room for embedding watermark. Currently, visible reversible watermarking is mainly focused raster image [4,10,12], few works has been done to vector graphics [9]. As for 3D mesh model, although some reversible watermarking schemes have been investigated in the past years, visible reversible watermarking has not been reported to the best of our knowledge.

In this paper, leveraging the elements and data characteristics of the 3D mesh model, a visible reversible watermark scheme for the 3D mesh model based on mesh subdivision is proposed to explicitly identify the copyright of model. The main contributions of the paper are summarized as follows.

(1) A visible reversible watermarking algorithm for 3D mesh model based on mesh subdivision is proposed. To the best of our knowledge, it is the first work on the visible reversible watermarking for 3D mesh model.
(2) Experimental results show that the watermarking algorithm has good robustness to translation, rotation, scaling, mesh refinement and mesh smoothing, the watermark is tightly integrated with the host, and the model will be destroyed or degraded if an unauthorized user tries to remove the watermark.

The rest of the paper is organized as follows. The related work is introduced in Sect. 2; the preprocessing of 3D mesh model is described in Sect. 3; The proposed scheme is depicted in Sect. 4; experimental results and analysis are provided in Sect. 5. Finally, some conclusions are drawn in Sect. 6.

2   Related Works

2.1   Reversible Watermarking for 3D Mesh Model

Jhou et al. proposed a reversible watermarking algorithm for 3D mesh model based on vertex histogram translation [6]. It used the last few digits of the distance between the vertex and the centroid of the model to construct a histogram, and then the watermark is embedded through the histogram translation. However, it requires the participation of the original model in watermark extraction, and it cannot against RST transformation. Chuang et al. built a histogram using the standard distance from the vertex to the centroid of the model, and then used the histogram translation technique to achieve reversible information hiding [3]. Since the standard distance is affine invariant, it is robust to

RST transformation, but it needs to record side information to complete the watermark extraction and carrier recovery process. Huang et al. calculated the distance between vertices according to the geometric similarity of the vertices in the neighborhood to construct a histogram, and realized the information embedding through histogram translation [5]. It effectively reduced the distortion of the watermark model, and is robust to vertex reordering operations, but its embedding capacity is small. Wu et al. proposed a 3D model reversible watermarking based on prediction error expansion [11]. It first obtains a set of independent vertices from the 3D mesh model, each of which uses the centroid of its neighboring vertices as prediction vertex, and then uses prediction error expansion to achieve watermark embedding. The watermark extraction and carrier recovery can be accomplished without the original model and other information, but the correlation of the vertices is not fully utilized, which leads to large distortion of the watermarked model more obvious. Zhu et al. proposed to select embeddable points based on the vertex curvature, and use the centroid coordinates of the vertex neighborhood as the coordinate prediction value of the embeddable points. Prediction error expansion is utilized to achieve information embedding [15]. It effectively improves the invisibility of the model, and it is robust to translation, rotation, and random noise, but its embedding capacity is limited. Zhang et al. proposed a multi-layer multiple embedding scheme using a hybrid strategy to predict errors [14]. It divides the area of the 3D mesh model into a uniform area and a non-uniform area, and different strategies are implemented to calculate the prediction error. The embedding strategy is adaptively chosen according to the geometric relationship between vertices and their neighboring vertices to realize information embedding. It can effectively reduce the distortion of the watermarked model, and multiple embedding is used to increase the capacity. Jiang et al. proposed a reversible information hiding using double-layer prediction errors [7]. It is a reversible data hiding (RDH) algorithm for 3D mesh models based on the optimal three-dimensional prediction-error histogram (PEH) modification with recursive construction coding (RCC). Firstly, it designs a double-layered prediction scheme to divide all vertices of 3D mesh model into the embedded set and the referenced set according to the odd-even property of indices in the vertex list. After that, the prediction errors(PEs) with a sharp histogram are obtained, and every three adjacent PEs are combined into one prediction error triplet (PET). A three-dimensional PEH with smaller entropy than one-dimensional PEH by utilizing the correlation among PEs is obtained. the three-dimensional PEH is projected onto one-dimensional space for scalar PET sequence, which is suitable for using RCC. Finally, the PET sequence and embed data are modified by RCC according to optimal probability transition matrix(OTPM). Experimental results show that it outperforms two state-of-the-art spatial-domain RDH algorithms for 3D mesh models.

2.2   Visible Watermarking for 3D Mesh Model

Compared with the reversible watermarking of the 3D mesh model, few studies have been done to visible watermarking. Lu et al. first proposed a visible watermarking for 3D mesh model [8]. Firstly, it projects the area to be embedded onto a two-dimensional plane, and then the watermark is embedded in the form of a vector. Finally, the area is projected back to the original model. The core is to divide the triangular facets into two parts inside and outside the watermark. The Sutherland-Hodgeman polygon cropping can achieve good effect on convex polygons. For concave polygons, it will be divided into several convex polygons, which increases the complexity. Based on this, An et al. proposed a visible watermarking algorithm for 3D mesh model based on mesh subdivision and boundary adaptation [1]. It first generates the watermark to be embedded through TTF character library, then selects the smoothest region of the model. The watermark is embedded in the smoothest region. Although these two algorithms achieve the visibility of the watermark, they cannot completely remove the watermark when it is needed, which affects the usability of the 3D model.

To achieve reversibility, this paper proposes a visible reversible watermarking scheme for 3D mesh model based on mesh subdivision based on literature [1,8]. It not only can explicitly identify the copyright of the 3D mesh model, but also can completely remove the watermark information to obtain the original 3D mesh model.

3   Preprocessing of 3D Mesh Model

To embed the visible watermark into 3D mesh model, some preprocessing operations need to be performed to 3D mesh model. They include the selection of the embedding area, projection, panning, zooming operations and etc. Given a 3D mesh model $G = (V, F)$, where V represents the vertex set of the 3D mesh model $V = \{v_i | v_i = (x_i, y_i, z_i) \in R^3, 1 \leq i \leq N\}$, F represents the triangular facets set that form the 3D mesh model $F = \{f_i = (v_p, v_q, v_r) \in R^3, 1 \leq i \leq N_f\}$. Here N represents the total number of vertices in the 3D mesh model, and N_f represents the total number of triangular facets in the 3D model. The embedded visible watermark W can be expressed as $W = (V^w, E^w)$, where V^w is the vertex set of watermark $V^w = \{v_i^w | v_i^w = (x_i^w, y_i^w), 1 \leq i \leq N^w\}$, E^w is the set of edges of the watermark $E^w = \{e_i^w = (v_i^k, v_k^w) | v_i^w, v_k^w \in V^w, 1 \leq i, k \leq N^w\}$, N^w is the number of vertices of the watermark.

3.1   Selection of Embedding Area

To achieve visibility of the embedded watermark, the smooth area on the 3D mesh model can be selected for watermark embedding. The smoothness of the 3D mesh model area can be measured by the sum of the angles between the adjacent facets of the vertices. By calculating the sum of the average angle of the normal vectors of triangular facets in the k-order neighborhood of the vertex v_i, the vertex with the smallest sum of average angles v_{imin} can be found, and it is determined as the smoothest point. It is defined by

$$D(v_i) = \sum_{n=p} d(v_i), \tag{1}$$

$$d(v_i) = \frac{\cos^{-1}(\overrightarrow{n_{f_k}}, \overrightarrow{n_{f_l}})}{N_i}, f_k, f_l \in N_f(v_i), \tag{2}$$

where p is the order of the neighborhood of the selected vertex, $d(v_i)$ is the sum of the average of the angles of the first-order neighborhood of the vertex, N_i is the number of first-order neighborhood triangles of vertex v_i, $N_f(v_i)$ is the first-order neighborhood facet set of vertex v_i, f_k, f_l are the triangular facets in the neighborhood of vertex v_i. $\overrightarrow{n_{f_k}}$, $\overrightarrow{n_{f_l}}$ are the normal vectors of f_k, f_l, respectively.

After selecting the smoothest point, to improve the calculation efficiency, the smoothest point v_{imin} is taken as the center of the circle, and a fixed threshold r is taken as radius, a smooth area S can be determined for watermark embedding. Figure 1 shows the first-order neighborhood of a vertex of the 3D mesh model.

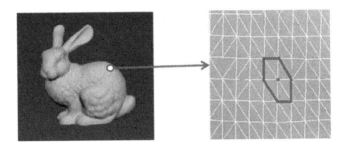

Fig. 1. The first-order neighborhood of a vertex

3.2   Projection of the Embedding Area

Since a two-dimensional watermark W is used to crop the mesh plane of the smooth area S, the smooth area S needs to be projected onto the two-dimensional plane. The essence of the projection is the transformation of the three-dimensional coordinate system [13]. Projection is to transform the direction of the original coordinate axis Z axis into the direction of the normal vector $\overrightarrow{n} = (x_n, y_n, z_n)$ of the smooth point v_{imin} under the condition that the origin is unchanged. The two-dimensional smooth area after the projection is recorded as $P = (V, F)$, and the set of vertices in this area is recorded as $V = \{v_i | v_i = (X_i, Y_i)\}$.

The coordinates of the watermark W and the smooth area P are both two-dimensional coordinates. Find the maximum and minimum of the horizontal and vertical coordinates of the watermark W and the area P. And recorded them as $x_{max}^w, x_{min}^w, y_{max}^w, y_{min}^w, X_{max}, X_{min}, Y_{max}, Y_{min}$, respectively. The coordinates of each vertex of the watermark W is multiplied by the scaling factor. So that $X_{max} - X_{min} > x_{max}^w - x_{min}^w, Y_{max} - Y_{min} > y_{max}^w - y_{min}^w$. After that, the center of gravity coordinates (x_0^w, y_0^w) of the watermark W is aligned with the center of gravity coordinates (X_0, Y_0) of the smooth area P. So that, $x_0^w = X_0, y_0^w = Y_0$. After these operations, the watermark W is all contained in the two-dimensional smooth area P. It guarantees that the watermark W can be completely displayed in the smooth area p.

4   The Proposed Visible Reversible Watermarking Scheme

The framework of the proposed scheme is shown in Fig. 2. It consists of four parts: visible watermark embedding, reversible watermark embedding, watermark removing, and recovery of 3D mesh model.

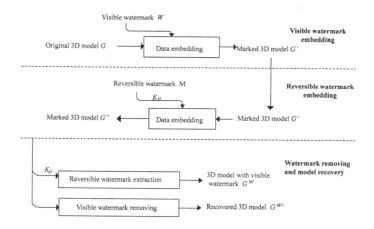

Fig. 2. The framework of the proposed scheme

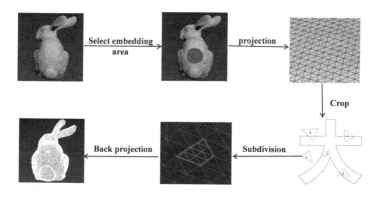

Fig. 3. The flow chart of visible watermark embedding

4.1   Watermark Embedding

For the given 3D mesh model $G = (V, F)$ and watermark $W = (V^w, E^w)$, the watermark embedding includes visible watermark embedding and reversible watermark embedding.

(1)Visible Watermark Embedding. The process of embedding a visible watermark is shown in Fig. 3. The specific process is as follows.

Step 1. The ray method is used to determine the positional relationship between all vertices in the projected two-dimensional smooth area P and the watermark W [2], and then whether the vertices in the model mesh are inside or outside the closed area of the watermark is determined.

Step 2. According to the positional relationship between the triangular facets in the projected two-dimensional smooth area P and the watermark W, the triangular facets which located inside the watermark W are subdivided [1].

Step 3. The two-dimensional plane with the watermark is projected back to the original 3D mesh model. Finally, the obtained 3D mesh model $G' = (V', F')$ is the model after embedding the watermark W. And the vertex set of the 3D model is $V' = \{v'_i | v'_i = (x'_i, y'_i, z'_i) \in R^3, 1 \leq i \leq N'\}$, the triangular facets set that forms the 3D mesh model is $F' = \left\{ f'_i = \left(v'_p, v'_q, v'_r\right) \in R^3, 1 \leq i \leq N'_f \right\}$, N' is the total number of vertices in the 3D model, and N'_f is the total number of the triangular facets in the 3D model.

(2)Reversible Watermark Embedding. To restore the original 3D mesh model, the index information of the visible watermark is embedded to the rest of the 3D model. The specific embedding process is as follows.

Step 1. Compare the vertices of the original 3D model G and the 3D model with visible watermark G', and record the different vertex sequences as $V^m = \{v_i^m | v_i^m = (x_i^m, y_i^m, z_i^m) \in R^3, 1 \leq i \leq N^m \}$, where N^m is the number of vertices. The vertex sequence is the visible watermark vertex sequence.

Step 2. Record the vertex index of the vertex sequence $V^m = \{v_i^m | v_i^m = (x_i^m, y_i^m, z_i^m) \in R^3, 1 \leq i \leq N^m\}$ as $M = \{m_i, 1 \leq i \leq N^m\}$, where N^m is the number of indexes in the sequence, and it is the same as the number of vertices in vertex sequence V^m.

Step 3. Convert the index m_i in the index sequence M into binary, and add "0" in the front of the binary to make all binary digits be the same. After that, a binary sequence $M' = \{m'_i, 1 \leq i \leq N^m\}$ is obtained.

Step 4. A stream encryption (such as RC4, etc.) is used to encrypt the sequence M' under the control of watermark embedding key K_H. A binary sequence M'_E is obtained.

Step 5. Embed M'_E into the 3D model $G' = (V', F')$, and then 3D mesh model $G'' = (V'', F'')$ with the visible reversible watermark is generated. The embedding is accomplished by the prediction error histogram modification proposed in [7]. It designed a double-layered prediction scheme to obtain prediction errors with a sharp histogram by utilizing the geometrical similarity among neighbouring vertices. It does not change the topology and keep the vertices for prediction unchanged in the embedding process, which guarantees the reversibility.

Step 6. Based on a key t, except the index contained in M, randomly generate two indices, and their corresponding vertices are represented as reference vertices v_{f1}, v_{f2}, respectively. The two reference vertices are kept as a secret.

4.2   Watermark Removal and 3D Mesh Model Recovery

The watermark removal and 3D mesh model recovery operations are performed to 3D mesh model $G'' = (V'', F'')$. The specific process is as follows.

Step 1. Obtain two reference vertices, and judge whether G'' is undergone RST operations according to the position relationship between v_{f1}, v_{f2}, and v'_{f1}, v'_{f2}. If it is, the corresponding inverse transform are performed to G''.

Step 2. Extract the embedded watermark sequence M'_E from the 3D mesh model G'' containing visible reversible watermarks, and then obtain the 3D mesh model $G^w = (V^w, F^w)$;

Step 3. Decrypt the sequence M'_E with information hiding key K_H to get watermark sequence M'.

Step 4. Remove the added "0" in front of the $m\prime_i$. So that the sequence $M\prime = \{m\prime_i, 1 \le i \le N^m\}$ is converted into $M = \{m_i, 1 \le i \le N^m\}$, and then get the vertex index of the embedded visible watermark;

Step 5. Traverse the 3D mesh model $G^w = (V^w, F^w)$, find the vertex index which is the same as the vertex index in the sequence $M = \{m_i, 1 \le i \le N^m\}$, and delete the corresponding vertex in the 3D model $G^w = (V^w, F^w)$ one by one. In this way, the removal of visible watermark is completed. Finally, the 3D mesh model $G^{w\prime} = (V^{w\prime}, F^{w\prime})$ is obtained.

5   Experimental Results and Analysis

5.1   Experimental Results

The configuration of the experimental platform is Intel(R)Core(TM)i5-5200U CPU @ 2.20 GHz, 8 GB RAM, Windows 8.1, Visual Studio 2017, and OpenGL. The 3D mesh models used in the experiment are shown in Fig. 4 (a)–(c), and the basic attributes are shown in Table 1.

Experiments are done to the above 3D mesh models, and the experimental results are shown in Fig. 4, where the selected embedding areas are shown in Fig. 4 (d)–(f), the watermarked 3D mesh models are shown in Fig. 4 (g)–(o), and the 3D mesh models after removing the visible watermark are shown in Fig. 4 (p)–(r). The attributes of the watermarked model can be found in Table 2. From the experimental results, the proposed visible reversible watermarking is effective.

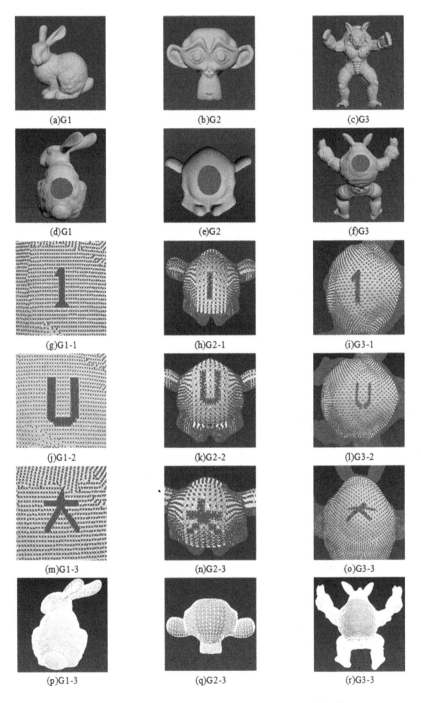

Fig. 4. Experimental results. (a)–(c):Original 3D mesh models;(d)–(f): Selected embedding areas of 3D mesh models; (g)–(o):3D mesh models with watermark; (p)–(r):3D mesh models after removing watermark.

Table 1. The basic information of experimental models

Graphics	Number of vertices (N)	Number of facets (F)
G1	34817	69630
G2	7830	15488
G3	106289	212574

Table 2. Basic information of experimental models after embedding watermark

Graphics	Number of vertices (N)	Number of facets (F)
G1-1	35078	70152
G1-2	35641	71278
G1-3	35191	70867
G2-1	7908	15648
G2-2	7957	15865
G2-3	7964	15760
G3-1	106407	212810
G3-2	106397	212790
G3-3	107122	218453

5.2   Performance Analysis

(1)Analysis of Reversibility. To quantitatively analyze the distortion of the
3D mesh model, the average moving distance $AvgD$ and the maximum moving distance $MaxD$ of all vertices are used as the measurement criteria. The
calculation is

$$AvgD = \frac{1}{M} \sum_{i=0}^{M-1} \sqrt{\sum_{j=1}^{3} \left(x_{ij} - x'_{ij}\right)^2}, \tag{3}$$

$$Max\,D = \max\left(\sqrt{\sum_{j=1}^{3}\left(x_{ij} - x'_{ij}\right)^2}\right), i \in \{0, 1, \ldots, M-1\}, \tag{4}$$

where M represents the number of vertices of 3D mesh model, x_{ij} represents
the jth coordinate of the ith vertices of the original 3D mesh model, and x'_{ij}
represents the jth coordinate of the ith vertices of the 3D mesh model after the
watermark is removed.

Experiments are done to analyze the difference between the original 3D mesh
model and the 3D mesh model after watermark removal, and the results are
listed in Table 3. It can be found that the average difference between the 3D
mesh model after watermark removal and the original 3D mesh model is less than
10^{-10}, while the maximum difference is less than 10^{-9}. The results show that the
method proposed by Jiang et al. outperforms two state-of-the-art spatial-domain

RDH algorithms for 3D mesh models. Furthermore, for the 3D mesh models after watermark removal, the number of vertices and the number of triangular facets are all the same as the original ones. Therefore, it can be concluded that the proposed scheme in this paper is reversible.

Table 3. Analysis of difference between recovery model and original model (10^{-10})

Graphics	Number of vertices (N)	Number of facets (F)	$AvgD$	$MaxD$
	After recovery	After recovery		
G1	34817	69630	6.1512	34.766
G2	7830	15488	3.2147	31.953
G3	106289	212574	9.4672	30.128

(2) Analysis of Robustness. For 3D mesh models, translation, rotation and scaling are typical operations in real applications. Robustness against these operations is very important for a reversible watermarking. Experiments were done to the watermarked models after different RST operations. Here, the robustness is evaluated by the correlation between watermark extracted after RST transformation and original watermark, and the results are listed in Table 4. From the results, the correlation between watermark extracted after RST transformation and original watermark are all 1.00. The main reason is that two reference vertices are selected after watermark embedding, and the host 3D mesh models can be restored according to two reference vertices even the watermarked 3D mesh models suffer RST operations. Therefore, it can achieve good robustness against RST operations.

Table 4. Correlation between watermark extracted after RST transformation and original watermark

Operations	G1	G2	G3
Rotation ($\rho = 30°$)	1.00	1.00	1.00
Rotation ($\rho = 60°$)	1.00	1.00	1.00
Rotation ($\rho = 120°$)	1.00	1.00	1.00
Translation (2.7, −1.2)	1.00	1.00	1.00
Translation (−6.3, 1.9)	1.00	1.00	1.00
Translation (−3.1, 1.3)	1.00	1.00	1.00
Scaling ($\zeta = 0.25$)	1.00	1.00	1.00
Scaling ($\zeta = 1.50$)	1.00	1.00	1.00
Scaling ($\zeta = 2.00$)	1.00	1.00	1.00

Experiments are also done to evaluate the influence of mesh subdivision and smoothing on the watermarked 3D mesh models, and the results are shown in Fig. 5. (a)–(c). From the results, it can be found that they have little influence on the visibility of the watermark.

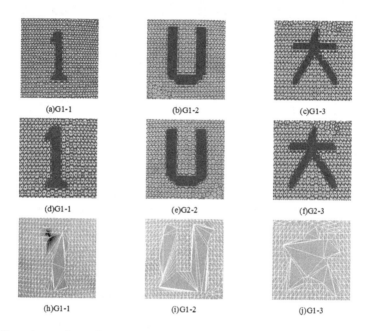

Fig. 5. Experimental results on the robustness of the 3D mesh model. (a)–(c): Subdivision attack; (d)–(f): Smooth attack; (g)–(o): The result of the attacker manually removing the watermark.

Meanwhile, analysis is made to analyze the possibility of unauthorized removal of the visible watermark, and the experimental results are shown in Fig. 5 (h)–(j). Since the adversary do not know the specific index position of the watermark vertex, they can only delete the watermark vertex one by one through the editing tool. Since there are thousands of embedded watermark vertices, it is difficult to delete all watermark vertices. In addition, the watermark vertices and the rest vertices of 3D mesh model are tightly integrated. In the process of deleting the watermark vertices, the rest of the 3D mesh model's vertices are inevitably deleted, which will destroy the 3D mesh model and the 3D mesh model is useless.

6 Conclusion

In this paper, a visible reversible watermarking for 3D mesh model based on mesh subdivision is proposed. It not only can explicitly identify the copyright of the 3D mesh model, but also the authorized users can completely remove the watermark to recover the original 3D mesh model. Experimental results show that the operations including translation, rotation, scaling, subdivision, smoothing, and etc. have no influence on the visibility of the embedded watermark. When an unauthorized user attempts to delete the watermark or steal the 3D mesh model, it will destroy the model and make the model unusable, because the visible watermark is closely integrated with the 3D mesh model. To the best of our knowledge, this is the first work of visible reversible watermarking for 3D mesh model. Our future work will be focused on the adaptive embedding of visible watermark.

References

1. An, X., Ni, R., Zhao, Y.: Visible watermarking for 3D models based on boundary adaptation and mesh subdivision. J. Appl. Sci. **34**(5), 503–514 (2016)
2. Chen, R., Zhou, J., Yu, L.: Fast method to determine spatial relationship between point and polygon. J. Xi'an Jiaotong Univ. **41**(1), 59–63 (2007)
3. Chuang, C.H., Cheng, C.W., Yen, Z.Y.: Reversible data hiding with affine invariance for 3D models. In: IET International Conference on Frontier Computing. Theory, Technologies and Applications, pp. 77–81 (2010)
4. Hu, Y., Jeon, B.: Reversible visible watermarking and lossless recovery of original images. IEEE Trans. Circ. Syst. Video Technol. **16**(11), 1423–1429 (2006)
5. Huang, Y.H., Tsai, Y.Y.: A reversible data hiding scheme for 3D polygonal models based on histogram shifting with high embedding capacity. 3D Res. **6**(2), 20 (2015)
6. Jhou, C.Y., Pan, J.S., Chou, D.: Reversible data hiding base on histogram shift for 3D vertex. In: Third International Conference on Intelligent Information Hiding and Multimedia Signal Processing (IIH-MSP 2007), vol. 1, pp. 365–370. IEEE (2007)
7. Jiang, R., Zhang, W., Hou, D., Wang, H., Yu, N.: Reversible data hiding for 3D mesh models with three-dimensional prediction-error histogram modification. Multimed. Tools Appl. **77**(5), 5263–5280 (2018)
8. Lu, C., Zhu, C., Wang, Y.: Visible watermarking for three-dimensional mesh model in two views. J. Image Graph. **19**(7), 1068–1073 (2014)
9. Peng, F., Ming, W., Zhang, X., Long, M.: A reversible visible watermarking for 2D cad engineering graphics based on graphics fusion. Signal Process.: Image Commun. **78**, 426–436 (2019)
10. Tsai, H.M., Chang, L.W.: A high secure reversible visible watermarking scheme. In: 2007 IEEE International Conference on Multimedia and Expo, pp. 2106–2109. IEEE (2007)
11. Wu, H.t., Dugelay, J.L.: Reversible watermarking of 3D mesh models by prediction-error expansion. In: 2008 IEEE 10th Workshop on Multimedia Signal Processing, pp. 797–802. IEEE (2008)
12. Yang, Y., Sun, X., Yang, H., Li, C.T., Xiao, R.: A contrast-sensitive reversible visible image watermarking technique. IEEE Trans. Circ. Syst. Video Technol. **19**(5), 656–667 (2009)
13. Zeng, W., Tao, B.: Non-linear adjustment model of three-dimensional coordinate transformation. Geomatics Inf. Sci. Wuhan Univ. **28**(5), 566–568 (2003)

14. Zhang, Q., Song, X., Wen, T., Fu, C.: Reversible data hiding for 3D mesh models with hybrid prediction and multilayer strategy. Multimed. Tools Appl. **78**(21), 29713–29729 (2019)
15. Zhu, A., Zhang, C., Yang, X., Gao, X.: Reversible watermarking of 3D mesh models using prediction-error expansion. In: 2010 3rd International Congress on Image and Signal Processing, vol. 3, pp. 1171–1175. IEEE (2010)

Multimedia Forensics

Defocused Image Splicing Localization by Distinguishing Multiple Cues between Raw Naturally Blur and Artificial Blur

Xiaoyu Zhao[1,2], Yakun Niu[1,2], Rongrong Ni[1,2(✉)], and Yao Zhao[1,2]

[1] Institute of Information Science, Beijing Jiaotong University, Beijing 100044, China
`rrni@bjtu.edu.cn`
[2] Beijing Key Laboratory of Advanced Information Science and Network Technology, Beijing 100044, China

Abstract. Splicing is one common type of forgery that maliciously changes the image contents. To make the forged image more realistic, blurring operations may be conducted to partial image regions or splicing edges to promise visual consistency. Revealing the blurring inconsistency among the whole image regions contributes to the splicing detection. However, for the defocused image already containing blur inconsistency, the existing methods cannot work well. Splicing detection and localization in defocused image is a challenging problem. In this paper, we overcome this problem by distinguishing multiple cues between raw naturally blur and artificial blur. Firstly, after the overlapped image blocks partition, three kinds of feature sets are extracted based on posterior probability map, noise histogram and derivative co-occurrence matrix. Then, an effective classifier is trained to determine the blur property of each pixel. Finally, a localization map refinement is proposed by fusing color segmentation probability map to improve the quality of the locating result. Experimental results demonstrate that the proposed method is very effective to detect splicing for the defocused images. The localization accuracy also outperforms the existing methods.

Keywords: Splicing localization · Raw naturally blur · Artificial blur

1 Introduction

Digital images can be easily tampered due to the availability and accessibility of diverse and powerful editors. Image tampering can falsify the truth and maliciously mislead the public. Therefore, detecting whether an image is the original output from the camera or tampered becomes increasingly attractive. Splicing is one common tampering operation that intends to change the image

This work was supported in part by the National Key Research and Development of China (2018YFC0807306), National NSF of China (U1936212, 61672090), and Beijing Fund-Municipal Education Commission Joint Project (KZ202010015023).

X. Zhao et al. (Eds.): IWDW 2020, LNCS 12617, pp. 153–167, 2021.
https://doi.org/10.1007/978-3-030-69449-4_12

contents. In the process of splicing, the cropping of image areas will generate dentate boundaries, and forgers often need to blur such boundaries in order to hide the tampering traces. When the sharpness of the splicing region is higher than the background of the original image to be spliced, forgers even blur the whole splicing region artificially. This is to ensure that the splicing region is visually consistent in sharpness with the background of the original image, so as to make the splicing image more realistic. And multiple images of different types of blur (motion blur or out-of-focus blur) can also be spliced together by forgers. Thus, the splicing detection through blur clues plays an important role in tampering detection.

In early studies, some methods were proposed to detect splicing by revealing the blurring inconsistency among image regions. Some authors proposed the method [1,3] to detect the incompatibility of different blur types in the image. For example, motion blur caused by camera shakes cannot only exist in the partial regions of the image, but exists globally. Therefore, the tampering traces can be detected in such a splicing image which the stationary objects (walls or buildings) contain motion blur and other image areas contain only out-of-focus blur. Other methods are mainly proposed based on that the local representation of blur are not consistent with actual information about image. These detection methods include detecting the inconsistency in the extent of blur [2–4,25], abnormal hue [19], and the direction and detail inconsistencies of motion blur [13,20,24]. However, for the defocused image already containing blur inconsistency, existing methods are difficult to identify the splice and accurately locate tampering regions.

For the defocused image, it already contains a certain extent of out-of-focus blur in the shooting process, which only depends on the own parameters of the camera and the distance of focus. In this paper, we refer to this blur when the camera parameters are fixed and only the focusing distance is controlled as raw naturally blur. The emergence and application of Single Lens Reflex (SLR) cameras make photos formed into different extents of raw naturally blur inevitably. Figure 1 (a) shows an original image with raw naturally blur formed during optical photography. When the forger wants to splice (a), he needs to apply an artificial blur consistent with background to the splicing area in order

(a) Original image   (b) Forged image

Fig. 1. (a) An original image with raw naturally blur. (b) A forged image generated by splicing an artificial blurred region in the original image (a).

to achieve a more pleasant visual effect, as shown in Fig. 1 (b). In essence, the dissimilarity between raw naturally blur and artificial blur can be employed as evidence of tampering, once it is discovered that can reveal the forgery history. While it is not taken into consideration in previous approaches. In this paper, we proposed a new method for splicing detection and localization by exploring the dissimilarity between raw naturally blur and artificial blur in defocused images.

In order to distinguish raw naturally blur from blurs caused by artificially manipulated, the blur types studied in this paper are explained in detail firstly. Raw naturally blur is specifically defined as an out-of-focus blur that exists in images stored directly in RAW format without any processing of the image pixels. It has not been affected by camera embedding operations. And it is a function of the distance between the subjects and a camera, which is different from lens blur in optical aberration. We define artificial blur as a manipulation of digital image pixels by some photo editing tools, and it is also similar to actual out-of-focus effects. Since out-of-focus blur can be classified into parametric and non-parametric in practice, then we discuss the research based on parametric types and extend our application to more complex non-parametric types. The parametric out-of-focus blur is symmetric and modeled as a cylinder disk with radius R. The non-parametric one is considered more complex and it could be an asymmetric shape. In several works [5, 12, 14], it has been confirmed that non-parametric out-of-focus blurs are closer to optical out-of-focus blur from the camera. In this paper, we study the parametric form of out-of-focus blur which are more realistic in image forensics applications, and then apply them to parametric and non-parametric blurry forgery localization. Based on analysis of imaging procedure, we find that raw naturally blur is captured directly in the presence of specific camera noise [17]. While artificial blur is the smoothing of image pixels after imaging. It can be considered as a de-noising operation performed by a filtering kernel. So we find the dissimilarity between the two in the image noise component. In addition, blurring affects the edge information of the image and the transition of the pixel value smoothly. Our purpose is to expose these dissimilarities between raw naturally blur and artificial blur by describing features from posterior probability map, noise histogram and derivative co-occurrence matrix.

In summary, the main contributions of our work are as follows. Firstly, multiple cues are explored for splicing regions detection in defocused image based on posterior probability map, noise histogram and derivative co-occurrence matrix. Compared with using single cue, the proposed multiple cues can reduce the misjudgment and improve the detection accuracy. Then, a localization map refinement process is designed based on the map fusion. It is not only makes full use of the binary result of SVM classification, but also image content characteristics, by regarding for the intention of the forger to conceal the spliced area. Experimental results demonstrate that the proposed method outperforms the existing methods for splicing detection as well as localization. Furthermore, the proposed method is robust to some post processing attacks, such as parametric/non-parametric bluring and JPEG compression.

2   Proposed Method

2.1   Investigating the Dissimilarity

As mentioned earlier, raw naturally blur is caused by different distances between the subjects and a camera. While we can obtain that artificial blur operates the selected area through moving the sliding filter to produce a smooth effect [27]. Filter functions of different window scales to make different blurry patterns, and the blurring intensity is controlled by parameters. Therefore, we extract features based on the unique pixel value smoothing, high-frequency noise reduction and edge information elimination effects generated by artificial blur which is different from raw naturally blur.

Posterior Probability Feature. The artificial blurring processing of a clear image I is modeled as follows:

$$B(i,j) = I(i,j) \otimes K(i,j) + \eta(i,j) \tag{1}$$

where B is an artificial blurry image, K is a point spread function (PSF) represented by a filter kernel, $\eta(i,j)$ is the additive white gaussian noise, such as $\eta(i,j) \sim \mathrm{N}(0,\sigma^2)$ and $\otimes$ is the convolution operation, respectively. From the perspective of the pixel values, which in raw naturally blur is determined by the shooting content and there is usually no regular linear relationship. Each pixel value $B(i,j)$ after artificial blurs can be approximately expressed by a linear combination of the pixel values of its neighborhoods with the same operation mode. The linear relation between $B(i,j)$ and their neighborhoods is given by:

$$B(i,j) = \sum_{x \neq 0}^{n} \sum_{y \neq 0}^{n} a_{x,y} B(i+x, j+y) - E \tag{2}$$

where $a_{x,y}$ is the weighting coefficient and n is the number of its neighbouring pixels. The coefficient is identical everywhere and controlled by the filter function only. It is verified artificial blurs can make the image pixels highly linear correlated with each other. E denotes the matrix composed of the error between the predicted and the actual gray value of each pixel, which is going to be $E = \sum_{x}^{n} \sum_{y}^{n} a_{x,y} B(i,j), a_{0,0} = -1$. Relearning this coefficient through images perhaps yield a little powerful effect, but it takes tremendous effort and is not very rewarding. Thus, based on the proper preset of $a_{x,y}$ from experience [7], we filter the input image and get the error matrix E. The gaussian transformation is then applied to each element of the error matrix E: $pMap = \exp(-E^2)$. During this transformation, big values of E are mapped to small values (near 0) of pMap in the range of $[0,1]$, and small of E are mapped to big (near 1) of pMap. The result map is the probability of error E between actual pixels value of input blurry image B and the expected pixel value of center position predicted by the adjacent pixels, that is the posterior probability map (pMap).

Because of the strong correlation between the pixels of artificial blurry image B, the element value of the error matrix E calculated will be so small that the probability value in the pMap will be large. However, it is opposite in raw naturally blurry image that the element value of E will be so large that the probability value in the pMap will be small. Which is caused by similar colors within the shooting scene of raw naturally blur or the pixel smoothing effect of artificial blur. Since the interference factors such as illumination of the shooting process, the pixel values of the same color are not necessarily linearly correlated. The color similarity of the shooting scene formed in raw naturally blurred image has limited influence on the posterior probability. Therefore, the large average of the posterior probability matrix is more likely due to the linear transformation from artificial blur of the pixel value. And we prove that the pMap values of artificial blurring images are closer to 1 (i.e. the posterior probability is larger), whereas the posterior probabilities of raw naturally blur images are usually small. We get the first feature on the pMap calculated above: $F_1 = mean(pMap)$.

Noise Histogram Feature. Due to the internal factors of the camera, raw naturally blur images obtained directly from the camera often contain the unknown actual noise pattern [9]. However, the purpose of artificial blur is to retain the low-frequency content information of the image and filter the high-frequency edged information, no matter what degrees of filter function. In order to achieve the image blurring effect, it is equivalent to low-pass filter, and obviously the high-frequency noise is usually reduced. Although the noise patterns from different camera sources can be used as cues for image classification, it only was effective when detecting image splicing region from different cameras. Thus, we aim to calculate statistical characteristics of the high-frequency noise histogram to expose the trace left by artificial blur in raw naturally blur.

The high-frequency noise residual image is denoted as $Noise$, which refers to the difference between the un-processing image and the artificially blurred image: $Noise(i, j) = I(i, j) - B(i, j)$. The histogram of high-frequency noise residual image can capture the changes of neighboring pixels. We denote the histogram of $Noise$ is,

$$hist(k) = \sum_i^M \sum_j^N \delta(Noise(i, j) = k) \tag{3}$$

where $\delta(\cdot)$ is the impulse function and the image size is $M \times N$. The value of $hist(k)$ represents the distribution frequency of the image gray level when the pixel value is k.

When we focus on the bar to the central value equals to 0 in the high-frequency noise histogram, it is the peak of histogram located. The higher peak value indicates the less high-frequency noise. It is obviously that artificial blur makes high-frequency noise less and the peak more prominent than raw naturally blur, while the value of the bar to the other central positions of the histogram decrease. In order to facilitate studying the distribution difference of raw naturally blur and artificial blur in the histogram of high-frequency noise residual image, we normalize this distribution of histogram in Fig. 2 through divided it by

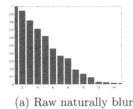

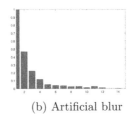

(a) Raw naturally blur   (b) Artificial blur

Fig. 2. The normalized histogram of high-frequency noise residuals in raw naturally blur (blue) and artificial blur (red). (Color figure online)

$hist(0)$ to enlarge the difference. To distinguish the difference of high-frequency component between raw naturally blur and artificial blur, the normalized distribution frequency by discarding the peak value $hist(0)$ is taken as the second feature: $F_2 = \frac{hist(k)}{hist(0)}$, with the value of k is a non-zero integer less than n.

Derivative Co-occurrence Matrix Feature One of the effects of blurring is the reduction of information at the edges of the image. In order to avoid the influence of input image contents on edge analysis, we investigate the change of the edges in the derivative domain. The derivative image is represented as:

$$\nabla B(i,j) = \nabla I(i,j) \otimes K(i,j) + constant \tag{4}$$

where $\nabla I(i,j)$ represents the derivative value of step edges in the input image, which is decreasing in an obvious intensity by blurring operations. And $\nabla B(i,j)$ represents the corresponding derivative value of step edges in the blurry image and $K(i,j)$ is same to the symbol K in Eq. (1). Compare Eq. (1) with Eq. (4), it effectively highlights the rate of the gray values changes and removes undesired distorted influences like additive noise $\eta(i,j)$ into constant.

The weakening of the edge components from blurring effects is shown in the derivative image as the decreases of the corresponding pixel values. We aim to obtain a performance similar to the gray level co-occurrence matrix, which reflects the local pattern of the image gray values about the adjacent interval and the change amplitude [22]. The joint probability of each pixel pair occurrence in derivative image is utilized to construct derivative co-occurrence matrix (DCM). It describes the spatial correlation of the neighbors of derivative pixels.

$$DCM(a,b) = \sum_{i_1,i_2}^{M} \sum_{j_1,j_2}^{N} \delta(\nabla I(i_1,j_1) = a, \nabla I(i_2,j_2) = b) \tag{5}$$

where (a,b) represents a pair of related pixels in a range of $[-T,T]$.

In addition, we have conducted the analysis deeply of how DCM shows the traces of artificial blur from raw naturally blur. We set the center pixel position of the derivative co-occurrence matrix to $(0,0)$ as the origin of coordinates, by establishing a planar cartesian coordinates system with it, the derivative co-occurrence matrix can be divided into four regions. We formulate R_i as:

$$R_i = \begin{cases} R_1, x > 0, y > 0 \\ R_2, x > 0, y < 0 \\ R_3, x < 0, y < 0 \\ R_4, x < 0, y > 0 \end{cases} \tag{6}$$

where the x-axis and y-axis are defined by the horizontal direction and vertical direction of DCM, respectively. $R_i(i = 1, 2, 3, 4)$ corresponds to the first, second, third, or fourth quadrant of the coordinate system is shown in Fig. 3.

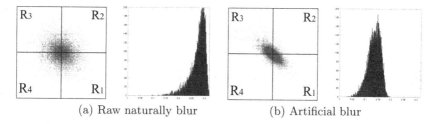

(a) Raw naturally blur   (b) Artificial blur

Fig. 3. The derivative co-occurrence matrix of raw naturally blur and artificial blur, respectively. And we graph the corresponding statistical result of the ratio of R_2, R_4 regions to the whole DCM for 10000 image blocks.

The value of the origin of DCM's coordinates represents the number of pixels with an image gradient of 0. Its increase depends on the smoothing effects. And noise pixels concentrate on the second and fourth quadrants of DCM, which points have opposite signs to the adjacent pixels. For raw naturally blurry image, its amount of noise is numerous and will not be cut without human processing. However, the noise in artificial blurry image will inevitably be reduced in part, making the ratio of R_2, R_4 regions smaller. It has been proved that artificial blur reduce the ratio of R_2, R_4 of the whole DCM, while there is no significant reduction in raw naturally blur. Here, we show this in the statistics results of 10000 image blocks in raw naturally blur and artificial blur as is shown in Fig. 3. And the derivative co-occurrence matrix (DCM) is taken as the third feature: $F_3 = DCM(a, b)$, its dimension is $(2T + 1)^2$.

2.2 Feature Classification

Based on the features of posterior probability, noise histogram and derivative co-occurrence matrix are all sensitive to smoothing effects of pixels, we incorporate the proposed three features to classify image blocks into raw naturally blur and artificial blur. When the scale of the block is smaller, the extracted features performance poorer. More time resources will be spent on training for classification learning and the training results will be relatively terrible because of extremely

Table 1. Comparison on the detection accuracy and training time loss of SVM classifier for images of different scales.

Scale (pixels)	16×16	32×32	64×64	96×96	128×128
Accuracy	74.12	84.06	**94.57**	95.12	97.77
Time loss	120.19	74.23	**62.31**	68.59	175.98

insufficient pixel. On the contrary, the larger the block size is, the stronger the performance expression of extracted features will be. In other words, if artificial blurring operation is applied to the whole space domain, our features will show better performance due to the increase of detection image size. However, it is time-consuming to extract the features of large patch, and it will lead to rough boundary localization in splicing detection. Therefore, according to the experimental comparison in Table 1, a moderate size of 64×64 was selected for our training. Since some image patches may not contain sufficient information to be reliably analyzed for forensics purposes [10]. That is, some smooth blocks without much effective information enough, which are not reliable in detection and have to be singled out. The common canny edge detector is used to obtain the number of edge pixels in an image block. If both the number of edge pixels and the variance larger than thresholds, the block is considered as a texture block. For a 64×64 block, the threshold of variance is set for 500 and the threshold of edge pixels is selected to be 1% of the total number of pixels in a block. Consequently, we designed this process for blocks of images with sufficient capacity to encode information, which to single out blocks that are not suitable for forensic SVM training.

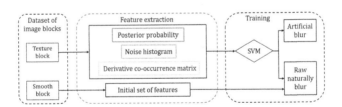

Fig. 4. SVM training to classification.

We adopt Support Vector Machines (SVM) with radial basis function (RBF) to achieve feature-based image blocks classification. The penalty parameter of the error term and the kernel parameter were selected by grid search method with cross validation [6], and the optimal parameters were trained on the whole input image block set. The SVM model training of classification is shown in Fig. 4. The training data are a large amount of texture block samples through cropping with their corresponding labels, which comes from images of multiple outdoor scenes. For smooth blocks, we assign an initial set of features and corresponding raw naturally blur labels to form its special samples. Specificly, the initial feature is defined in an all 0 vector and has the same dimension as the feature vector of texture blocks. This is to forcibly assign the specified label to the annotated smooth block, and the accurate judgment of the test image will not be affected in the subsequent experiments in this way.

2.3 Refine Splicing Localization

In practical forensic applications, we think more interests in forged regions localization than image-level detection. When the input image was possibly spliced, we expect to find the forged area perfectly and output the pixel level localization result. First, the overlapping partition is carried out on the test image in order to get the input blocks required by SVM. Corresponding to the settings in SVM training and classification, smooth blocks in the overlapped blocks should be marked their indexes and assigned with raw naturally blur labels. Then, the trained SVM model is applied on texture blocks prediction. Second, the overlapping partition method can assign multiple labels for each pixel via predicted result when the preset of overlapping step size to 1. When each pixel is owned by different blocks, it contains n^2 labels where $n \times n$ is the block scale. It is worth noting that although smooth blocks are considered by SVM as raw naturally blur labels directly, there is little influence on the predicting outcomes. The reason is that our binary output map depends on a majority of votes strategy, that is, the label of each pixel in binary output is indicated by the main one of n^2 overlapping labels.

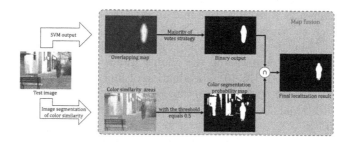

Fig. 5. The scheme of refining localization by map fusing.

Third, in order to refine the localization result, we aim to correct labels in binary output. The scheme of refining localization by map fusing is shown in Fig. 5 specifically. It can be used as the basis of our refinement of localization that forgers tend to choose color-independent objects for splicing to mask discontinuous splicing backgrounds. We utilize Otsu's method to segment color similarity areas of the input color image, and calculate its posterior probability matrix (see Sect. 2.1 for calculation details). Then, we traverse each local 8×8 pixel area of the binary output of SVM. Due to artificial blur makes posterior probability (p) relatively larger than raw naturally blur. So we retain those areas of pMap to a suspicious area for artificial blur as our color segmentation probability map, when pixels of $p > 0.5$ more than half of the each local 8×8 pixel area. Finally, the intersection of the color segmentation probability map and the SVM binary output makes the boundary of binary output more refined and we get the final localization result.

3   Experiments

3.1   Experimental Setup

To generate raw naturally blurred images, we shot images of the same 50 scenes at different focusing distances by using the manual focusing. While keeping the aperture and lens focal length constant and ensuring maximum stability, we used Nikon D600 SLR camera to build our TIF image dataset with photo size of 6016×4016 pixels. For our localization experiments, we select the image patches with size of ranging from 512×512 to 1664×1216 pixels. Because the depth of field is proportional to the lens focal length, aperture value, and the square of focusing distance [26], we control the former two variables so that the cause of raw naturally blur was only due to the manual focusing controlled by the user.

Then, the un-processed dataset is categorized into two parts of high and low blur degrees. Both two datasets stored in TIF format as well as JPG formats for the compression quality factors of 100, 90, 80. Where the image blocks with high degrees of out-of-focus blur are directly served as raw naturally blur without any other operations. And artificial blur is obtained by filtering the image blocks with low degrees of blur. A lower bound of artificial blur degrees is chosen as $R = 2$, where the blur degree is negligible [3]. The upper bounds of blur degrees are chosen differently since it is required to pay attention to keeping its quality indices much the same to raw naturally blur to eliminate the effects of the inconsistency of blur degrees. In the process of this, the corresponding relationship between the blur degrees and the quality indices of patches is established through the two methods [15,16] of evaluation of image quality indices. Finally, 5000 raw naturally and 5000 artificial blur patches are used to form the training set, and the additional 3000 patches were randomly selected respectively to form the testing set.

Since we believe the above research based on the dissimilarity between raw naturally blur and artificial blur is universal for both parametric and non-parametric blur, we verify the feature effectiveness through experiments under the same conditions. We first applied a cylinder disk with radius R as the filter function in parametric blurring tampering. Then we created 20 categories of asymmetric kernel and resize into several scales for different blurring degrees to generate non-parametric blurring kernel.

By randomly choosing 800 raw naturally blur images, we splice each of them with a local extrinsic region from other images. Then, we artificially blurred the spliced regions by setting equal to the blur extent of each background image area. We obtain 800 images each using parametric/non-parametric artificial blur kernel, a total of 1600 images for splicing detection. Each category consists of

200 splicing images with tampering area proportions of 5, 15, 25 and 35 percents. Through the analysis and attempts of the proposed feature, we also assign specific values to the relevant parameters. The range of k is set at $n = 5$ for $1 \leq k \leq n$ in the noise histogram and the number of a pair of relating pixels (a, b) is set at $T = 3$ in DCM, respectively. For an input image with R, G and B color spaces, by transforming into the YCbCr domain, the values of the three channels that luminance (Y), blue (Cb) and red (Cr) components are obtained. Therefore, we take the luminance (Y) component as the input image and extract the 55-dimensional feature to SVM for the following experiments.

3.2   Result Evaluation

Our first experiment examines the validity of the proposed feature for classification of raw naturally blur and artificial blur. For image filtering detection, we do comparative analysis with TPM feature from the state-of-the-art method Subramanyam et al. [21] on raw naturally blur and artificial blur image blocks.

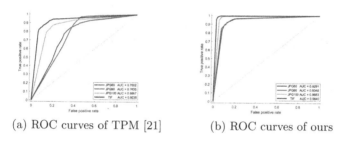

(a) ROC curves of TPM [21]   (b) ROC curves of ours

Fig. 6. Comparison with TPM [21] method on ROC curves of features.

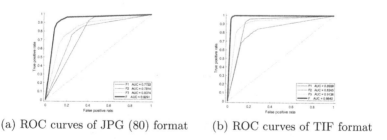

(a) ROC curves of JPG (80) format   (b) ROC curves of TIF format

Fig. 7. Comparison with each single feature on ROC curves for ablation experiment.

As is shown in Fig. 6, our feature is highly effective against classification between two categories of blocks (raw naturally blur and artificial blur) but the performance of TPM is obviously worse than ours. Especially in TIF format and JPG format at the compression factors equal and greater than 80, the AUC of us is still greater than 0.92. It shows that the proposed feature is more robust to JPEG compression. Meanwhile, as the features we proposed contain multiple cues, an ablation study is necessary to prove the effectiveness of each feature. Separately, we utilize the feature F_1, F_2 and F_3 to conduct image block classification experiments to obtain the corresponding three ROC curves. In Fig. 7, ROC curves of different image storage formats are summarized and compared, among which the thickest is the feature F our proposed ($F = [F_1, F_2, F_3]$). Figure 7 showed that when a single feature is used, the more mediocre effect than features connected by same weights. Consequently, we use connected multiple cues to the optimal AUC value.

Table 2. With confidence interval of 70%, our comparison of the recall rate of the CFA [8], JNB [23], NOI [18] methods in parametric and non-parametric data, respectively.

Parametic	CFA [8](%)	JNB [23](%)	NOI [18](%)	Ours(%)
JPG (80)	1.00	76.75	84.75	**94.38**
JPG (90)	37.75	82.00	93.88	**97.25**
JPG (100)	85.50	84.50	89.75	**99.00**
TIF	86.50	86.63	94.25	**99.13**
Non-parametic	CFA [8](%)	JNB [23](%)	NOI [18](%)	Ours(%)
JPG (80)	0.55	81.38	84.88	**95.13**
JPG (90)	33.25	83.00	90.00	**95.63**
JPG (100)	89.75	84.63	81.88	**96.38**
TIF	91.50	85.13	83.13	**94.25**

Next, we compare the performance of our method with Color Filter Array (CFA) [8], Just Noticeable Blur (JNB) [23], Noise Variance (NOI) [18] methods by considering various tampering area ratios in splicing images. In the first part, we calculate the coincidence rate of each result and corresponding ground truth as the evaluation indicator, the 70% confidence interval (i.e., count it as correct detection one when the coincidence rate is greater than 70%) is chosen. For test results less than this confidence interval, the area of misdetermination is large enough to make us skeptical of the whole. So the test results with more than 70% coincidence rate can be accepted and trusted. We derive the recall rate compared with other three methods as shown in Table 2. The CFA method is able to locate the tampering area well in TIF format and JPG format with the quality factor of 100, but greatly affected when the compression quality factor is 90 and almost fails when 80. And we all got the highest recall rate for each image storage format under the parametic/non-parametic artificial blur kernel.

In order to make a fair comparison of detecting results, it is necessary to binarize output artifacts from methods CFA [8], JNB [23], NOI [18], which show the probability that splicing forgery occurs. In the second part, we select its threshold by maximizing the average accuracy of raw naturally and artificial blur types detection for each method. If a probability value is larger than its threshold, it belongs to the spliced region and indicated by white and the reverse

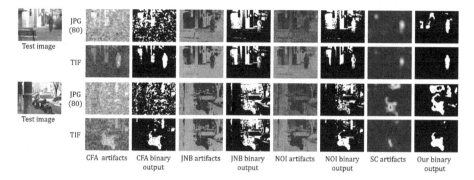

Fig. 8. Comparison of the locating results of test image in JPG (80) and TIF storage format with CFA [8], JNB [23], NOI [18] and SC [11] methods. The test image is at the beginning of the line, where the region inside the red curve is the splicing region. (The SC [11] method has only a qualitative result without its binary map.) (Color figure online)

Table 3. Comparison of locating accuracy of parametric and non-parametric dataset with other methods.

	Tamper Ratio	JPG (80)				JPG (90)				JPG (100)				TIF			
		CFA[8] (%)	JNB[23] (%)	NOI[18] (%)	Ours (%)	CFA[8] (%)	JNB[23] (%)	NOI[18] (%)	Ours (%)	CFA[8] (%)	JNB[23] (%)	NOI[18] (%)	Ours (%)	CFA[8] (%)	JNB[23] (%)	NOI[18] (%)	Ours (%)
Parametric	5%	59.43	75.68	78.71	83.76	71.73	80.19	83.82	83.66	76.08	82.15	84.66	89.46	76.93	85.36	85.95	89.07
	15%	55.45	76.72	79.68	87.62	67.62	79.87	85.32	88.41	79.31	81.81	85.92	92.96	80.00	81.99	87.28	92.81
	25%	55.96	77.52	80.63	86.04	69.45	80.90	87.00	88.41	80.22	82.74	86.59	91.48	81.17	82.91	88.30	91.37
	35%	56.61	74.60	78.71	87.78	68.97	76.94	84.47	89.72	81.35	78.36	84.54	94.01	82.41	78.61	86.07	93.89
Non-parametric	5%	54.92	76.56	79.15	83.12	66.92	79.11	82.46	81.43	82.61	80.57	80.47	84.92	83.40	80.77	81.23	84.06
	15%	55.33	77.61	80.32	87.46	66.97	80.33	83.95	87.35	82.91	81.69	81.99	89.88	84.21	81.89	82.56	88.90
	25%	55.86	78.41	81.26	85.44	68.27	81.42	85.41	86.81	82.60	82.72	81.77	88.92	83.74	82.89	82.67	88.12
	35%	55.72	77.56	80.42	89.30	67.28	80.43	84.53	90.05	82.88	81.77	81.89	92.34	83.90	81.97	82.65	91.53

by black. We also compared with a novel self-supervised method based on deep learning, which done for splicing localization by detecting image Self-Consistency (SC) [11]. The artifacts and binary output generated by the methods of CFA [8], JNB [23], NOI [18], SC [11] and ours for two example images in JPG (80) and TIF storage formats are shown in Fig. 8, respectively. It is to mention that instead of dealing with binary output result of the SC [11] method for calculating its accuracies, we did a qualitative analysis with it. Because it has a poor locating result on our forged dataset as well as a long time loss for detecting.

The locating accuracy is calculated by the coincidence rate between the binary result and the ground truth, and as is shown in Table 3. By comparing experiments on images of different storage formats, the CFA [8] method significantly reduces the detection accuracy when the compression factor is small, while JNB [23], NOI [18] and our method only change in a small range. It also verified the robustness of our proposed method to the JPEG compression attacks. And in the cases of different tampering ratios, only the detection accuracy of ours increases with the tamper area getting larger. Although other methods are not subject to the ratios of tamper area, our accuracy is outstanding than them. The result shows that previous works have low performance because they do not take the dissimilarity we propose between raw naturally blur and artificial blur into account. Moreover, our proposed feature works well even for complicated forms of non-parametric blur kernels and JPEG compression post-processing attacks.

4 Conclusion

In this paper, we solve a challenging issue of splicing detection and localization in defocused image by distinguishing raw naturally blur and artificial blur. The first contribution is the extraction of three kinds of feature sets based on posterior probability, noise histogram and derivative co-occurrence matrix though detailed and extensive analysis. Secondly, an effective classifier of our multiple cues is trained to determine the blur property of small image blocks and we achieved outstanding performance of classification in comparative experiments of filtering detection. Finally, we refine the localization map of splicing in defocused image by fusing with the color segmentation probability map. Moreover, the proposed method is robust to parametric and non-parametric blurring and JPEG compression. Experimental results demonstrate that the proposed method is superior to state-of-the art methods in splicing detection as well as the localization accuracy. Still, we consider the more work ahead by deeply exploring the contribution to each feature of different weights to blurring forensics.

References

1. Bahrami, K., Kot, A.C.: Image splicing localization based on blur type inconsistency. In: 2015 IEEE International Symposium on Circuits and Systems (ISCAS), pp. 1042–1045 (2015)
2. Bahrami, K., Kot, A.C., Fan, J.: Splicing detection in out-of-focus blurred images. In: 2013 IEEE International Workshop on Information Forensics and Security (WIFS), pp. 144–149 (2013)
3. Bahrami, K., Kot, A.C., Li, L., Li, H.: Blurred image splicing localization by exposing blur type inconsistency. IEEE Trans. Inf. Forensics Secur. 10(5), 999–1009 (2015)
4. Cao, G., Zhao, Y., Ni, R.: Edge-based blur metric for tamper detection. J. Inf. Hiding Multimed. Signal Process. 1(1), 20–27 (2010)
5. Chakrabarti, A., Zickler, T., Freeman, W.T.: Analyzing spatially-varying blur. In: 2010 IEEE Computer Society Conference on Computer Vision and Pattern Recognition, pp. 2512–2519 (2010)
6. Chang, C.C., Lin, C.J.: LIBSVM: a library for support vector machines. ACM Trans. Intell. Syst. Technol. (TIST) 2(3), 1–27 (2011)
7. Dong, W., Wang, J.: Jpeg compression forensics against resizing. In: 2016 IEEE Trustcom/BigDataSE/ISPA, pp. 1001–1007 (2016)
8. Ferrara, P., Bianchi, T., De Rosa, A., Piva, A.: Image forgery localization via fine-grained analysis of CFA artifacts. IEEE Trans. Inf. Forensics Secur. 7(5), 1566–1577 (2012)
9. Gou, H., Swaminathan, A., Wu, M.: Intrinsic sensor noise features for forensic analysis on scanners and scanned images. IEEE Trans. Inf. Forensics Secur. 4(3), 476–491 (2009)

10. Güera, D., Zhu, F., Yarlagadda, S.K., Tubaro, S., Bestagini, P., Delp, E.J.: Reliability map estimation for CNN-based camera model attribution. In: 2018 IEEE Winter Conference on Applications of Computer Vision (WACV), pp. 964–973 (2018)
11. Huh, M., Liu, A., Owens, A., Efros, A.A.: Fighting fake news: image splice detection via learned self-consistency. In: Ferrari, V., Hebert, M., Sminchisescu, C., Weiss, Y. (eds.) ECCV 2018. LNCS, vol. 11215, pp. 106–124. Springer, Cham (2018). https://doi.org/10.1007/978-3-030-01252-6_7
12. Jang, J., Yun, J.D., Yang, S.: Modeling non-stationary asymmetric lens blur by normal Sinh-Arcsinh model. IEEE Trans. Image Process. **25**(5), 2184–2195 (2016)
13. Kakar, P., Sudha, N., Ser, W.: Exposing digital image forgeries by detecting discrepancies in motion blur. IEEE Trans. Multimed. **13**(3), 443–452 (2011)
14. Kee, E., Paris, S., Chen, S., Wang, J.: Modeling and removing spatially-varying optical blur. In: 2011 IEEE International Conference on Computational Photography (ICCP), pp. 1–8 (2011)
15. Li, L., Dong, W., Wu, J., Li, H., Lin, W., Kot, A.C.: Image sharpness assessment by sparse representation. IEEE Trans. Multimed. **18**(6), 1085–1097 (2016)
16. Li, L., Lin, W., Wang, X., Yang, G., Bahrami, K., Kot, A.C.: No-reference image blur assessment based on discrete orthogonal moments. IEEE Trans. Cybern. **46**(1), 39–50 (2016)
17. Lukas, J., Fridrich, J., Goljan, M.: Digital camera identification from sensor pattern noise. IEEE Trans. Inf. Forensics Secur. **1**(2), 205–214 (2006)
18. Mahdian, B., Saic, S.: Using noise inconsistencies for blind image forensics. Image Vision Comput. **27**(10), 1497–1503 (2009)
19. Peng, F., Wang, X.l.: Digital image forgery forensics by using blur estimation and abnormal hue detection. In: 2010 Symposium on Photonics and Optoelectronics, pp. 1–4. IEEE (2010)
20. Rao, M.P., Rajagopalan, A.N., Seetharaman, G.: Harnessing motion blur to unveil splicing. IEEE Trans. Inf. Forensics Secur. **9**(4), 583–595 (2014)
21. Ravi, H., Subramanyam, A.V., Emmanuel, S.: Forensic analysis of linear and non-linear image filtering using quantization noise. ACM Trans. Multimed. Comput. Commun. Appl. **12**(3), 1–23 (2016)
22. Sang, Q.B., Li, C.F., Wu, X.J.: No-reference blurred image quality assessment based on gray level co-occurrence matrix. Pattern Recognit. Artif. Intell. **26**(5), 492–497 (2013)
23. Shi, J., Xu, L., Jia, J.: Just noticeable defocus blur detection and estimation. In: Proceedings of the IEEE Conference on Computer Vision and Pattern Recognition, pp. 657–665 (2015)
24. Song, C., Zeng, P., Wang, Z., Li, T., Shen, L.: Image forgery detection based on motion blur estimated using convolutional neural network. IEEE Sensors J. **19**(23), 11601–11611 (2019)
25. Uliyan, D.M., Jalab, H.A., Wahab, A.W.A., Shivakumara, P., Sadeghi, S.: A novel forged blurred region detection system for image forensic applications. Exp. Syst. Appl. **64**, 1–10 (2016)
26. Xiao, H., et al.: Defocus blur detection based on multiscale SVD fusion in gradient domain. J. Visual Commun. Image Representation **59**, 52–61 (2019)
27. Zhou, L., Wang, D., Guo, Y., Zhang, J.: Blur detection of digital forgery using mathematical morphology. In: Proceedings of the 1st KES International Symposium on Agent and Multi-Agent Systems: Technologies and Applications, pp. 990–998 (2007)

Deepfake Video Detection
Using Audio-Visual Consistency

Yewei Gu[1,2], Xianfeng Zhao[1,2], Chen Gong[1,2], and Xiaowei Yi[1,2(✉)]

[1] State Key Laboratory of Information Security, Institute of Information Engineering, Chinese Academy of Sciences, Beijing 100093, China
[2] School of Cyber Security, University of Chinese Academy of Sciences, Beijing 100049, China
{guyewei,zhaoxianfeng,gongchen,yixiaowei}@iie.ac.cn

Abstract. Benefit from significant advances in deep learning, widespread Deepfake videos with the convincing manipulations, have posted a serious of threats to public security, thus the identification of fake videos has become increasingly active in current researches. However, most present Deepfake detection methods concentrate on exposing facial defects through direct facial feature analysis while merely considering the synergies with authentic behavior information outside the facial regions. Meanwhile, schemes based on meticulous-designed neural networks are rarely efficient to provide subjective interpretations of the final identification evidences. Therefore, to further enrich the diversity of detection method and increase the interpretability of detection evidences, this paper proposes a self-referential method to exploit audio-visual consistency by introducing synchronous audio recordings as reference. In preprocess phase, we propose an audio-visual matching strategy based on phonemes to segment videos, and control experiments have proved that strategy outperforms common equal-length partition. To deal with such video segments, an audio-visual coupling model (AVCM) is employed for audio-visual feature representations, then similarity metrics are measured for mouth frames and related speech segments. Actually, synchronized pairs mean the high scores of similarity and asynchronous pairs opposite. The evaluations on DeepfakeVidTIMIT indicate that our method has achieved competitive results compared with current main methods, especially in high quality datasets.

Keywords: Deepfake detection · Audio-visual consistency · Audio-visual matching · Convolutional neural networks

1 Introduction

Recently, tampered video content has a great improvement in the realism of forgery by employing advanced neural networks, especially the generative adversarial networks (GANs). Deepfake videos in which faces are swapped with

This work was supported by NSFC under 61902391, National Key Technology R&D Program under 2019QY2202 and 2019QY(Y)0207, and IIE CAS Climbing Program.

© Springer Nature Switzerland AG 2021
X. Zhao et al. (Eds.): IWDW 2020, LNCS 12617, pp. 168–180, 2021.
https://doi.org/10.1007/978-3-030-69449-4_13

another one's faces are almost indistinguishable from human vision nowadays. Widespread abuse of the accessible open face-swapped techniques has led to mass fake news appearing in our life, which poses a significant concern to national security and individual reputations. To contend with this growing threats, the continual research efforts on Deepfake detection are in urgent demand.

Traditional forgery videos are processed per frame by using picture editing tools, like Adobe Premiere, and discordant visual defects caused by manual manipulations are easily identified by the human eyes. Comparatively, Deepfake videos are visually apparent, which achieve a higher level of realism. Therefore, the current approaches are concerned not only with single frame, but also with inconsistency of character behaviors (e.g. blink [11], head pose [19], etc.). In terms of accuracy, most present methods have reached over 90% of accuracy on open datasets. However, compared to the intuitive judgment by human eyes, few main methods can provide persuasive evidences assisting operators to make subjective inferences. To provide an interpretable solution, considering authentic behavior information outside facial regions, it might be a well attempt to explore the consistency of speeches and lip movements. In the case of audio as reference, the related lip features can be easier to identify.

Audio-visual synchronisation is a crucial problem in film and television industries. In recent years, researches on consistency detection have conducted through audio-visual content analysis. Different from Deepfake videos, the detection objects of consistency algorithms are real videos with shifted audio. In comparison, Deepfake videos with similar imitations are less likely to be recognized although existing some visual flaws in details. The work [7] has revealed the low efficient performances of the lip-sync detection method [8] on Deepfake videos, which indicates that present consistency algorithms need significant improvements to adapt to Deepfake detection.

Despite the efforts on single frame processing, face-swapped techniques seldom reach a high level on inter-frame processing. Therefore, exploring the consistency of diversiform character behaviors has been a noteworthy solution. As we consider, for a speaker, lip movements are the most delicate and ingenious behaviors compared with other facial behaviors, which are hardly to be meticulously imitated. To achieve self-reference, our work is dedicated to audio-visual consistency detection by introducing synchronous audio as reference. Our work is an improvement above the current audio-visual consistency algorithms. In terms of partition strategy, we employ a matching strategy based on phonemes instead of common equal-length partition. The varied-length audio-visual segments provide more elaborate and generic representations of features across individuals. To adapt the matching strategy, we also propose an audio-visual coupling model (AVCM) based on convolutional neural networks (CNNs) to capture the correlation of lip shapes and speeches. We develop various experiments with different strategies on DeepfakeTIMIT [7] and present favorable comparisons with previous works. The results indicate the audio-visual consistency detection on Deepfake videos is significantly effective.

2   Related Work

2.1   Deepfake Detection

There are two critical techniques—face synthesis and facial region splicing introduced in Deepfake production pipelines which profoundly affect the quality of the synthetic videos. Detections [10,11,13] on visual flaws in the facial regions have been active areas. Li et al. [11] observed that the realistically-looking fake faces were lack of eye blinking. The work [10] discovered that generated images based on deep neural networks have distinct differences from real images in specific color spaces. Missing reflections and deficient details in the eye and teeth regions were exposed in [13]. Additionally, some recent researches [12,19] have exploited flaws caused by facial region splicing. Yang et al. [19] have presented the inconsistency of 3D head poses between central facial regions and facial splicing regions by using facial landmarks. The work in [12] explored the face-warped artifacts surrounding facial regions. Succeed in previous works, we focus on synthetic mouth regions to explore the consistency of speaking.

2.2   Audio-Visual Consistency Detection

Traditionally, linguistic knowledge considered as important domain knowledge are extensively utilized to solve audio-visual consistency problems, such as [9,14]. Phoneme recognition and classify were applied in [9] to confirm precise audio-visual correspondence, yet feature representations were in low efficiency. In recent years, the development of neural networks has brought new solutions [2,8,18]. Siamese networks were employed for audio-visual feature representations [2]. Furthermore, Torfi et al. [2,18] proposed 3D convolutional neural networks for feature extraction in the time domain and space domain. The work in [8] linearly integrated audio features and visual features as inputs of long short-term memory (LSTM) network to evaluate speaking consistency while it was proved inefficient on Deepfake detection [7]. We suppose researchers who benefit from massive data tend to overlook hidden linguistic knowledge behind speeches. In our work, phonemes are introduced into matching strategy and a multimodal coupling model is employed for similarity measure of audio-visual segments.

3   Workflow

The method accepts videos with authentic audio as inputs. We process these inputs in four main stages in workflow (Fig. 1). In the phoneme alignment stage (Sect. 3.1), audio is aligned and clipped into segments in sequence at the phoneme level. In the mouth extraction stage (Sect. 3.2), mouth regions are captured. Then we have audio segments and mouth frames matched into pairs by employing specific partition strategies in audio-visual matching stage (Sect. 3.3). At last, in similarity measure stage (Sect. 3.4), AVCM is trained using these pairs to capture the relationship between audio and video features, then distinguishing positive samples from negative ones by measuring similarity of audio-visual embedding.

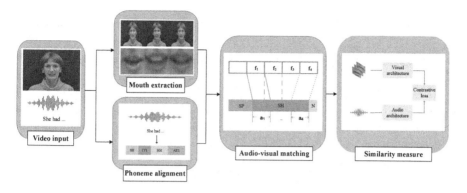

Fig. 1. Method overview. In the preprocessing phase (phoneme alignment, mouth extraction and audio-visual matching), audio segments and mouth frames are matched into pairs. In the model training phase (similarity measure), samples are divided according to calculated similarity

3.1 Phoneme Alignment

Phonemes are objective physical phenomenons which distinguish one word from another as the basic phonetic units. Actually, one phoneme corresponds to one specific vocal action. The acoustic features of a specific phoneme have fair commonality across individuals. For a given audio, phoneme alignment refers to computing the identity and timing of phonemes in sequence. In this section, we have audio aligned by using a phoneme-based alignment tool P2FA [16] which eventually produces an ordered phoneme sequence as follow.

$$A = (c_1, c_2, ..., c_n) \tag{1}$$

$$c_k = (label, st, et) \tag{2}$$

We denote A as the given audio which consists of n phoneme intervals c_k, and each c_k represents a specific phoneme interval in which the phonetic identity (*label*), start time (*st*) and end time (*et*) are recorded.

3.2 Mouth Extraction

We use face landmarks detected by dlib library to locate mouth regions. The points numbered 48~67 are used to mark the mouth area, in which point 66 is located near the center of mouth area. The point is used as the center of the clipped rectangle. Each frame is clipped into a 64×40 ($W \times H$) gray-scale mouth image to avoid the influence of colors.

3.3 Audio-Visual Matching

Unlike common equal-length partition strategy, we aim to produce audio-visual pairs by employing a new matching strategy and have video clipped base on

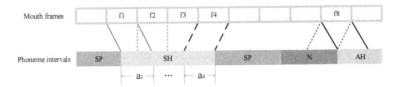

Fig. 2. Audio-visual matching. Each frame with a specific phoneme interval is matched to a fixed-length audio segment. A frame such as f_5 cross over two phoneme intervals are reuse. SP means a pause in speech.

phonemes. For a phoneme interval c_i, as mouth frames are numbered in sequence, we locate its corresponding frame sequence as Eq. (3), in which f_s and f_e represent the start frame number and end number of c_i respectively. $A[\cdot, \cdot]$ means a speech clip and f represents the video frame rate. In our design, each mouth frame f_i in located frame sequence and its corresponding audio segment a_i make up an audio-visual pair. The sequential pairs with same phoneme labels are assembled into groups, called phoneme units.

$$a_i = \begin{cases} A\left[st, st + len\right], & i = f_s \\ A\left[(i-1) * len, i * len\right], & i \in (f_s, f_e) \\ A\left[et - len, et\right], & i = f_e \end{cases} \tag{3}$$

$$len = 1/f \tag{4}$$
$$f_s = \lceil st * f \rceil \tag{5}$$
$$f_e = \lceil et * f \rceil \tag{6}$$

As Fig. 2 shows, audio is clipped into continuous phoneme intervals, such as SH, N, AH, etc. For a phoneme interval c_i, it covers several mouth frames and each frame f_i is related to a fixed-length audio segment a_i. This process doesn't introduce extra audio features beyond c_k and pauses (SP in Fig. 2) in audio are effectively skipped.

$$P_i = (f_i, a_i) \tag{7}$$
$$U_i = (P_{i1}, P_{i2}, ..., P_{in}) \tag{8}$$
$$V_i = (U_{i1}, U_{i2}, ..., U_{im}) \tag{9}$$

Input videos are split into sequential phoneme units. The mathematical expressions for these audio-visual clips are described in Eq. (7, 8, 9). A audio-visual pair is recorded as P_i. Meanwhile, U_i refers to a phoneme unit which consists of n P_is and V_i represents a video composed by m U_is.

3.4 Similarity Measure

In this section, we propose AVCM, which is a CNN model consists of audio architecture and video architecture. AVCM is employed to measure similarity

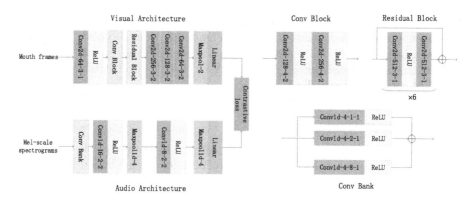

Fig. 3. The architecture of AVCM. The parameters of convolution layers are recorded as Conv-channel-kernel-stride and pooling layers are recorded as Pool-kernel.

of audio-visual pairs. The similarity is a metric used to indicate the degrees of synchronization of audio-visual pairs. Ideally, synchronized pairs correspond to high scores and asynchronous pairs opposite. The model is trained to expand the similarity of positive samples and minify the negative ones. More details are described in Sect. 4.

4   AVCM

4.1   Architecture

The details of AVCM are described in Fig. 3. Different from Siamese Networks in [18], we have developed two specific convolutional networks for audio embedding and visual embedding respectively. In visual architecture, $Conv2d$ layers are used to process lip shape features. Considering the complicated and changeable characteristics of lips, we improve the depth of the architecture. To solve the problems of gradient diffusion and network degradation caused by deep networks, $Residual\ Block$ based on ResNet [6] is introduced to ensure the stable convergence of the model. $Maxpool$ is used for downsampling and reducing redundant data. In audio architecture, $Conv1d$ layers are employed processing the mel-scale frequency information, and $Conv\ Bank$ which is made up of serial convolutional blocks with successively increasing receptive fields, is adapted to capture long-term semantic information.

4.2   Audio-Visual Features

Image Features. A phoneme unit U_i with variable-quantity pairs is used as a training batch. In each pair, mouth frame is converted to an array with 64×40 ($W \times H$) dimensions, then is normalized by subtracting mean and dividing standard deviation.

Audio Features. We use mel-scale spectrograms as audio features. In our experiments, mel-scale spectrograms are computed from a power spectrum (power of magnitude of 2048-sized STFT) on 40-ms windows length. To maintain the same scale with image features, the mel-scale spectrograms are then processed with same normalization.

4.3   Contrastive Loss

To optimize the coupling of audio-visual features, a contrastive loss initially proposed in [5] is employed. The loss function updates the model parameters by reducing distances of synchronous pairs and increasing distances of asynchronous pairs. The expressions are described as Eq. (10).

$$E = \frac{1}{2n} \sum_{i=1}^{n} yd^2 + (1 - y)\mathbf{max}(Margin - d, 0)^2 \qquad (10)$$

$$d = \|f_i - a_i\|_2 \qquad (11)$$

In Eq. (10), E represents the loss of U_i with n pairs. $Margin$ is a predefined parameter and y is a flag argument that $y = 1$ represents synchronous pairs and $y = 0$ represents opposite. The Euclidean distance d in Eq. (11) is calculated for each P_i, which is negatively correlated with similarity metrics.

4.4   Model Details

Generally, a fixed value is set for batch size. While the pairs in U_i are variable-quantity, to fit training batchs produced by matching strategy, a batch size adaptation is employed by resetting the batch size to quantity of pairs for each input. In the proposed model, we utilize the Adam optimizer for gradient descent as initializing *learningrate* to 0.0005. To bring a more steady convergence, *weight decay* is set to 0.0001 and *grad norm* is set to 5. *Margin* in contrastive loss is set to 700–900. Further details can be found in our implementation code: https://github.com/BrightGu/AVCDetection.

After training, real and fake samples are effectively divided into different camps, in which real samples corresponding to higher similarity metrics while fake samples opposite. Then, an optimal partition value is selected as threshold used to determine whether a video is true or false in inferring stage.

5   Experiments

5.1   Dataset

To facilitate the development of Deepfake detection technologies, researchers have published varied types of Deepfake datasets, including Faceforensics++ [15], Celeb-deepfakeforensics [20], DFDC [1], etc. However, no synchronized voices are provided in most present datasets, which leaves our method with no implementation conditions. Therefore, we evaluate proposed method on Vid-TIMIT [17] and DeepfakeTIMIT [7] where synchronous audio-visual pairs are

Table 1. Data distributions of train set and test set

DataSet	Type	Train	Test
VidTIMIT [17]	Video	350	80
	Clip	11277	2592
DeepfakeTIMIT [7]	Video	240	80
	Clip	7721	2592

produced from VidTIMIT and asynchronous pairs from DeepfakeTIMIT. Videos in DeepfakeTIMIT are derived from VidTIMIT with faces swapped using the open source GAN-based approach, and transcripts in speeches are from TIMIT corpus [4]. There are two subsets in DeepfakeTIMIT, referred to as high quality (HQ) with 128×128 input/output size and low quality (LQ) with 64×64 size. The detections are conducted on videos and phoneme units (clips in Table 1) respectively. More details are showed in Table 1. In the following text, we use DeepfakeTIMIT uniformly to represent the two datasets.

5.2 Evaluation on DeepfakeTIMIT

The experiments are conducted on two quality of subsets: HQ and LQ, with two different evaluate objects: phoneme units (clips in Fig. 4) and videos. The similarity metric of a clip is the average metric of pairs in the clip, and the metric of a video is the average metric of clips in the video. The score distributions and the FAR curves are described in Fig. 4. On the whole, the real samples and fake samples are both distributed in a concentrated manner although there are some overlaps which can't be separated. For the detections on same object (Fig.4(a, c) or Fig. 4(b, d)), the performances on LQ are slightly better than HQ. In terms of the sample distributions, fake samples are in smoother distributions in high-score areas on LQ, which means the number of indistinguishable extreme points has been reduced. The possible explanation is that compared with LQ, the image quality of HQ are closer to real samples which eventually leads to a lower-efficient divide on HQ. For the detections on same datasets (Fig. 4(a, b) or Fig. 4(c, d)), the performances on videos surpass the clips. We consider the phenomenon is related to the mean-value strategy, which eliminates the effects of extreme values to some extent. As the Fig. 4(b, d) shows, the overlaps are reduced, and the smoother distributions of videos may probably caused by the mean-value strategy.

From Fig. 4(a, c) or Fig. 4(b, d), we can see the similar trends of the FAR curves. The similar trends mean that the inner distributions of samples in LQ and HQ are the same. Actually, despite the input/output sizes in generative model are different, the architectures remain the same, which eventually lead to the similar trends. Moreover, in the case of consistent data distributions, LQ with lower image quality is easier to identify than HQ.

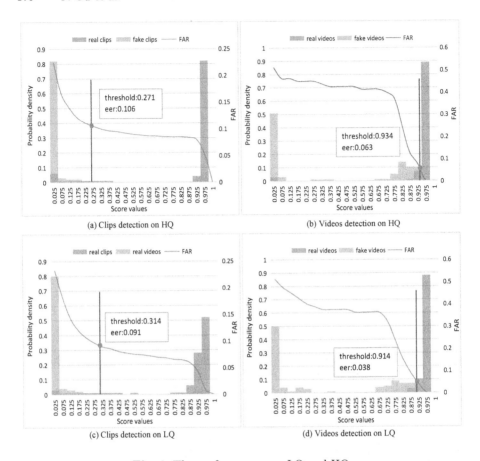

(a) Clips detection on HQ

(b) Videos detection on HQ

(c) Clips detection on LQ

(d) Videos detection on LQ

Fig. 4. The performances on LQ and HQ.

Table 2. Comparison of two matching strategies

Dataset	Unit type	AUC (%)		EER (%)		FRR@FAR10% (%)	
		PBP	ELP	PBP	ELP	PBP	ELP
LQ	Video	**99.20**	97.05	**3.75**	12.50	**1.00**	14.00
	Clip	**95.64**	92.80	**9.10**	11.86	**8.00**	16.00
HQ	Video	**97.44**	95.05	**6.25**	11.25	**5.00**	15.00
	Clip	**94.09**	91.43	**10.57**	18.49	**11.00**	27.00

5.3 Comparison of Two Matching Strategies

To evaluate our method on Deepfake detection, the primary step is to find an effective audio-visual matching strategy to have video segmented. In our method, we have video segmented based on phonemes, called Phoneme-Based Partition (PBP). As reference, we introduce another anchor strategy by clipping video into equal-length segments, called Equal-Length Partition (ELP). During training, the only difference of the two strategies is batch size. For PBP, a batch means a phoneme unit which consists of variable-quantity pairs, so we conduct extra adaptive strategy.

Table 3. Comparison on DeepfakeTIMIT

Dataset	Method	AUC (%)	EER (%)	FRR@FAR10% (%)
LQ	LSTM lip-sync [8]	/	41.8	81.67
	IQM+SVM [3]	/	**3.33**	**0.95**
	Face-Warp-ResNet50 [12]	**99.9**	/	/
	AVCM (ours)	99.2	3.75	1.00
HQ	IQM+SVM [3]	/	8.97	9.05
	Face-Warp-ResNet50 [12]	93.2	/	/
	AVCM (ours)	**97.44**	**6.25**	**5.00**

The experimental results are described in Table 2. From Table 2, PBP outperforms ELP both on LQ and HQ. For video detection, the AUC metrics indicate PBP has achieved 2.39% promotions on HQ and 2.15% promotions on LQ. Compared with ELP, PBP is more in line with the rules of speech division. Despite the extra work involved in phoneme alignment, PBP has realized promotions on accuracy eventually. Experiments show that the phoneme-based partition scheme can obtain more effective and universal audio-visual representations, and then achieves better performances. Subjectively, LQ is more likely to be exposed, thus the promotions (the proportion by which PBP exceeds ELP) on LQ, may have certain increases over HQ. However, the results haven't confirmed it, which indicate that promotions have no appreciable expansion although evaluate objects are in different qualities. We assume the possible reason is that the promotions are related to sample distributions, and has no relevance to the image quality. Indeed, the impacts based on different matching strategies are influenced by the sample distributions.

In terms of data preprocessing, ELP performs more commonly, and it doesn't require participation of transcripts. Relatively, the implementations for PBP seem more strict due to the extra work on phoneme alignment stage. On the whole, PBP remains slight promotions in accuracy, yet ELP is also considered as a valid alternative when there are no speeches or transcripts provided.

5.4   Comparison with Previous Methods

In this section, we choose three detection methods with different strategies in contrast to our method: a common audio-visual consistency detection method (LSTM lip-sync [8]), an image quality measurement method (IQM+SVM [3]) and a Deepfake detection method (Face-Warp-ResNet50 [12]). By contrast, we have observed Deepfake videos from three different perspectives (consistency of behavior, image quality and warpped artifacts) which reveals some problems in present Deepfake videos and further proves the feasibility of our method.

LSTM lip-sync [8] is a method used to checking audio-visual consistency on videos with shifted audio. The results in Table 3 indicate that the performance of it on Deepfake detection is distinctly undesirability. We consider the main problem is the unreasonable features it captures. The method concentrates on the study of continuous lip-movement information based on LSTM, yet such long-range information are universally considered hard to characterize which eventually lead to unsatisfactory results. Instead, our method explores the correlation of speech features and lip shapes at the frame level, which leads us to focus on not only lip shapes but also image quality. IQM+SVM [3] is a method based on image quality measurement with SVM classifier. The performances on HQ and LQ indicate that visual defects exist objectively in Deepfake videos and can be effectively identified, and the flaws in LQ are more pronounced than HQ. Moreover, without neural networks, traditional methods can also solve problems. Face-Warp-ResNet50 [12] is an effective Deepfake detection method verified by experiments, which aims to exposing distinctive artifacts around marginal areas of fake faces. The method presents the best score on LQ and a slightly decreasing score on HQ.

Compared with the above methods, the results in Table 3 demonstrate that our method has achieved the best performances on HQ. Nevertheless, the performances on LQ slightly lags behind IQM+SVM and Face-Warp-ResNet50. The explanations, as we consider, are that when image quality is in low degrees, IQM+SVM and Face-Warp-ResNet50 are more sensitive to obvious visual defects. While image quality is in high degrees, visual defects are relatively insidious. Compared with other facial areas, intricate lip shapes are harder to imitate, which eventually results in relatively distinct flaws. Therefore, our method which pays attention on mouth regions and introduces speech as reference, eventually achieves better performs on HQ. The results confirm that our method provides greater accuracy and feasibility.

Despite the imperfect of present forging technologies, we consider, along with the deeper researches on generative networks, the visual defects will be more insidious, and then better detectors will be in urgent need. In addition to the accuracy of the detectors remains to be improved, there also exist growing demands for multi-dimensional forensics on current Deepfake detection. The main reason is that it seems hard for people to make judgments based on unexplainable conclusions given by deep neural network models. It is generally expected to access evidence chains or semantic cooperations from multi-class detections, which requires that detection methods must be diversified.

6   Conclusion

In this paper, we present a Deepfake detection method by exploiting audio-visual consistency. As demonstrated, our approach explores visual defects in synthetic mouth regions by introducing authentic audio as reference. Additionally, the experiments indicate that phoneme-base partition performs a more efficient adaption than equal-length partition. Meanwhile, the method offers a new solution for forgery detection through a referential pattern. Compared with the existing methods that exploring unexplained visual features, our method greatly enhances the interpretability and richness of evidences and can assist operators to make subjective judgments to some extent. Based on varied experiments, we form a more precise understanding of audio-video consistency, especially introducing phonemes into matching strategies. Considering present vast excellent works on lip-sync video generation, we expect new breakthroughs on research of audio-video consistency, which will firmly play a import role in the fields of detection and generation.

References

1. Dolhansky, B., et al.: The Deepfake detection challenge dataset (2020)
2. Chung, J.S., Zisserman, A.: Out of time: automated lip sync in the wild. In: Chen, C.-S., Lu, J., Ma, K.-K. (eds.) ACCV 2016. LNCS, vol. 10117, pp. 251–263. Springer, Cham (2017). https://doi.org/10.1007/978-3-319-54427-4_19
3. Galbally, J., Marcel, S.: Face anti-spoofing based on general image quality assessment. In: 2014 22nd International Conference on Pattern Recognition, pp. 1173–1178 (2014)
4. Garofolo, J.S.: Timit acoustic phonetic continuous speech corpus. Linguist. Data Consort. **1993** (1993)
5. Hadsell, R., Chopra, S., LeCun, Y.: Dimensionality reduction by learning an invariant mapping. In: 2006 IEEE Computer Society Conference on Computer Vision and Pattern Recognition (CVPR 2006), vol. 2, pp. 1735–1742. IEEE (2006)
6. He, K., Zhang, X., Ren, S., Sun, J.: Deep residual learning for image recognition. In: Proceedings of the IEEE Conference on Computer Vision and Pattern Recognition, pp. 770–778 (2016)
7. Korshunov, P., Marcel, S.: Deepfakes: a new threat to face recognition? Assessment and detection. arXiv preprint arXiv:1812.08685 (2018)
8. Korshunov, P., Marcel, S.: Speaker inconsistency detection in tampered video. In: 2018 26th European Signal Processing Conference (EUSIPCO), pp. 2375–2379. IEEE (2018)
9. Lewis, J.: Automated lip-sync: background and techniques. J. Vis. Comput. Animat. **2**(4), 118–122 (1991)
10. Li, H., Li, B., Tan, S., Huang, J.: Detection of deep network generated images using disparities in color components. arXiv preprint arXiv:1808.07276 (2018)
11. Li, Y., Chang, M.C., Lyu, S.: In Ictu Oculi: exposing AI generated fake face videos by detecting eye blinking. arXiv preprint arXiv:1806.02877 (2018)
12. Li, Y., Lyu, S.: Exposing Deepfake videos by detecting face warping artifacts. arXiv preprint arXiv:1811.00656 (2018)
13. Matern, F., Riess, C., Stamminger, M.: Exploiting visual artifacts to expose Deepfakes and face manipulations. In: 2019 IEEE Winter Applications of Computer Vision Workshops (WACVW), pp. 83–92. IEEE (2019)
14. Morishima, S., Ogata, S., Murai, K., Nakamura, S.: Audio-visual speech translation with automatic lip syncqronization and face tracking based on 3-D head model. In: 2002 IEEE International Conference on Acoustics, Speech, and Signal Processing, vol. 2, pp. II-2117. IEEE (2002)

15. Rossler, A., Cozzolino, D., Verdoliva, L., Riess, C., Thies, J., Nießner, M.: Face-forensics++: learning to detect manipulated facial images. In: Proceedings of the IEEE International Conference on Computer Vision, pp. 1–11 (2019)
16. Rubin, S., Berthouzoz, F., Mysore, G.J., Li, W., Agrawala, M.: Content-based tools for editing audio stories. In: Proceedings of the 26th Annual ACM Symposium on User Interface Software and Technology, pp. 113–122. ACM (2013)
17. Sanderson, C.: The VidTIMIT database. Tech. rep., IDIAP (2002)
18. Torfi, A., Iranmanesh, S.M., Nasrabadi, N., Dawson, J.: 3D convolutional neural networks for cross audio-visual matching recognition. IEEE Access 5, 22081–22091 (2017)
19. Yang, X., Li, Y., Lyu, S.: Exposing deep fakes using inconsistent head poses. In: ICASSP 2019–2019 IEEE International Conference on Acoustics, Speech and Signal Processing (ICASSP), pp. 8261–8265. IEEE (2019)
20. Li, Y., Yang, X., Sun, P., Qi, H., Lyu, S.: Celeb-DF: a large-scale challenging dataset for Deepfake forensics. In: IEEE Conference on Computer Vision and Patten Recognition (CVPR) (2020)

A Machine Learning Approach to Approximate the Age of a Digital Image

Robert Jöchl$^{(\boxtimes)}$ ⓘ and Andreas Uhl ⓘ

Department of Computer Science, University of Salzburg, Salzburg, Austria
{robert.joechl,andreas.uhl}@sbg.ac.at

Abstract. In-field image sensor defects develop almost continually over a camera's lifetime. Since these defects accumulate over time, a forensic analyst can approximate the age of an image under investigation based on the defects present. In this context, the temporal accuracy of the approximation is bounded by the different defect onset times. Thus, the approximation of the image age based on in-field sensor defects can be regarded as a multi-class classification problem. In this paper, we propose to utilize two well-known machine learning techniques (i.e. a Naive Bayes Classifier and a Support Vector Machine) to solve this problem. The accuracy of each technique is empirically evaluated by conducting several experiments, and the results are compared to the current state-of-the art in this field. In addition, the prediction results are assessed individually for each class.

Keywords: Digital image forensics · In-field sensor defects · Image age approximation · Machine learning

1 Introduction

During the image acquisition processing pipeline, all the steps involved leave unique traces in the image. These traces are at the core of digital image forensics, as they can be used, for example, to detect image forgery or to identify the camera used. There are three main sources that leave traces during image acquisition (i.e. the lens, sensor, and color filter array) and their combination forms the acquisition fingerprint. For a comprehensive overview of techniques that exploit these traces, see [12].

In this paper, we focus on the traces introduced by the image sensor. Every image sensor has imperfections. Some of these occur during the manufacturing process, while others develop over a camera's lifetime. In [7], Jessica Fridrich shows how these imperfections can be used in digital image forensics. The Photo Response Non Uniformity (PRNU) of a sensor represents the variations in quantum efficiency among pixels. These variations result, among other reasons, from

X. Zhao et al. (Eds.): IWDW 2020, LNCS 12617, pp. 181–195, 2021.
https://doi.org/10.1007/978-3-030-69449-4_14

the slightly different pixel dimensions (i.e. due to imperfections in the manufacturing process). Since these variations are different for each device, the PRNU represents a sensor fingerprint and identifies the imager used [7].

In-field sensor defects arise after the manufacturing process. In principle, these defects are due to cosmic radiation, and once a pixel is defective, it can no longer heal. As a result, in-field sensor defects accumulate over time and by detecting their presence, the age of an image can be inferred. Image age approximation is relevant to forensic analysts when the chronological order among pieces of evidence can help to deduce a causal relationship between events. Since the time information from the EXIF header is usually not trustworthy, reliable methods for image age approximation are required.

Fridrich et al. introduced a technique to approximate the age of an image based on the presence of in-field sensor defects in [8]. The assumed situation is that a forensic analyst is provided with a set of trusted images for which the acquisition time is known and a second untrustworthy set. The goal is to approximate the age of the images from the second set relative to the first trustworthy set. To achieve this, the authors proposed a maximum likelihood principle to estimate the defect parameters, to define their onset times and to approximate the acquisition times. This approach is currently the only technique that uses sensor defects for temporal image forensics.

In [8], the authors estimated a certain index of the trustworthy set as the age of an image under investigation. We remind that image age approximation is based on in-field sensor defects, and therefore only the time interval between two consecutive defect onset times can be predicted. In this paper, we consider image age approximation as a multi-class classification problem and propose to utilize traditional machine learning techniques to solve it. In particular, we train a probabilistic (i.e. a Naive Bayes Classifier (NB)) and a geometric (i.e. a Support Vector Machine (SVM)) classifier. The classifier accuracy of each approach is empirically evaluated by conducting several experiments based on three data sets, and the results are compared to the current state-of-the-art in this field (introduced in [8]).

The remainder of this paper is organized as follows: In the next section we discuss previous work on in-field sensor defects and image age approximation. The different machine learning techniques are described in Sect. 3, followed by a discussion of the practical implementation issues. The conducted experiments and results are described in Sects. 5 and 6. Finally, the key insights are summarized in Sect. 7.

2   Related Work

2.1   In-field Sensor Defects

As the name suggests, in-field sensor defects develop in-field, after the manufacturing process. The characteristics of these defects are studied in [5,9–11,15,16].

In principle, a pixel is made of silicon and cosmic rays can damage this silicon layer. Typically, less than 2.3% of the pixel's area are affected by this damage. As a result, a defect only affects the behavior of a single pixel. Nevertheless, because of preprocessing steps like demosaicing (i.e. interpolation), the defect spreads over its neighboring pixels. Since cosmic radiation is the defect source, defective pixels develop independently of one another and are uniformly distributed over the entire image sensor. For example, defects would occur in clumps if environmental stress were the defect source. In addition, if the sensor is stored at the same altitude (i.e. with similar exposure to cosmic radiation), new defects develop almost continuously. Chapman et al. analyze the influence of ISO expansion, pixel size decrease, and smaller sensor areas on the defect development rate in [2]. This combination increases the defect development rate significantly.

The value of a pixel is determined by the incoming light. In particular, the photoelectric effect is used to convert the incoming light into electrons, and by quantizing the resulting voltage, the pixel's value is defined. The quantization setting defines the pixel's dynamic range (usually $[0, 255] \in \mathbb{N}$). A defect reduces this range by introducing an offset. To model a defect we rely on the definition in [8], except we define the model for a single pixel (i.e. all variables are scalars instead of matrices). In this context, let I be the intensity of the incoming light to which a certain pixel is exposed. The function $f(I) \, \mathbb{R} \to [0, 255] \in \mathbb{N}$ maps the incoming light to the pixel's dynamic range. A defective pixel can be defined as

$$f(I) = I + IK + \tau D + c + \Xi, \tag{1}$$

where K represents the PRNU factor. The dark-current D depends on the ISO setting, the exposure time and the temperature. These factors are combined in τ and multiplied with D. Thus, the offset of the pixel varies as τ varies. A fixed offset is denoted by c, and Ξ represents all other noise. Based on the defect type, either D, c or both parameters are high (i.e. a hot pixel has a high dark-current). We remind that the PRNU K develops during the manufacturing process and therefore does not contain any age information. Examples of such in-field sensor defects are illustrated in Fig. 1. The right image shows defects extracted from a dark-field image. Dark-field images are calibration images where the camera's shutter is closed (i.e. $I \approx 0$) and therefore the defect or rather its offset becomes visible. As illustrated, a defect can vary in shape, size, and the affected color channel. If the defect offset is high enough, the defect becomes visible in regular scenes, as shown in Fig. 1 (left).

2.2   Method Proposed by Fridrich et al.

In [8], the authors assume that an analyst is given two sets of images. One set contains trusted images in chronological order and the second set contains images where the capturing time is unknown. To approximate the age of an image from the second set relative to the first trustworthy set, the authors propose a maximum likelihood technique.

In particular, a defect is considered being noise, and by applying a denoising filter (i.e. a median filter) the defect is filtered out. The median filter residual, which is obtained by subtracting the median filtered image from the original image, thus contains the defect's magnitude (i.e. $IK + \tau D + c$). In [8], the

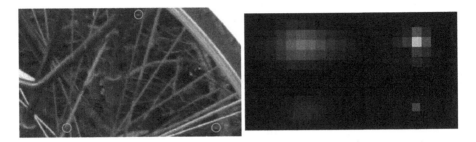

Fig. 1. Examples of different in-field sensor defects extracted from a regular scene image (left) and from captured dark-field images (right).

authors assume that the difference between the noise residual and the sum of all defect parameters (K, D, c) is normally distributed, with a mean of zero and a variance σ^2. To estimate the unknown parameters $\theta = (K, D, c, \sigma)$ (before and after the defect onset) and the defect onset time j, the authors propose a maximum likelihood approach. The onset time j represents the index of the trusted set where the defect onset happens, and $\hat{\theta}^{(0)}$ denotes the estimated parameter before the onset and $\hat{\theta}^{(1)}$ afterwards. To approximate the age of an image under investigation, a maximum likelihood approach is used again, i.e.

$$\hat{j} = \operatorname*{argmax}_{j} \prod_{i \in \Omega} \frac{1}{\sqrt{2\pi}\hat{\sigma}_{(i)}^{(\Psi)}} \exp \frac{W_{(i)} - \left(I_{(i)}K_{(i)}^{(\Psi)} + \tau D_{(i)}^{(\Psi)} + c_{(i)}^{(\Psi)}\right)}{2\hat{\sigma}_{(i)}^{2(\Psi)}}, \qquad (2)$$

where Ω denotes the set of all defective pixels and $W_{(i)}$ is the residual value of the ith defective pixel from that image. Dependent on Ψ the defect parameters are regarded either before or after the defect onset, i.e. $\Psi = 1$ if j is greater as the estimated onset time for the ith defective pixel. Hence, the resulting approximated acquisition time $\hat{j}$ is the index at which the difference between the residual value and the sum of the defect parameters is minimum for all defects.

3   Machine Learning Approach

In this paper, we assume the same situation as in [8], where S denotes the set of trusted images. However, instead of predicting a certain index of S, we consider image age approximation as a multi-class classification task. This is reasonable because image age approximation is based on the presence of in-field sensor defects, and, therefore only the time interval between two consecutive defect onsets can be predicted. Hence, the temporal prediction accuracy is bounded by these intervals. A common way to solve such classification problems is by using machine learning techniques. For this purpose, we propose to utilize a probabilistic and a geometric classifier. However, the features used must be defined before applying them.

3.1   Features

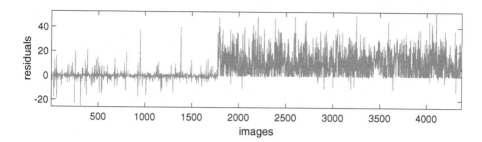

Fig. 2. Residuals of a defective pixel over the set of chronologically ordered images. The defect onset at about image 1770 is clearly visible.

The spatial defect property can be described as a single peak in the middle of a smooth image area (i.e. an area with constant pixel values). Such a peak can be smoothed out by applying a median filter. At the positions at which the median filter responds, the median filter residuals are not equal to zero. Furthermore, the residual value reflects the height of the peak, or rather the defect's magnitude. For this reason the median filter residual is a good defect indicator and is used as classification feature. In Fig. 2, the residual values of a single defective pixel over S are shown. It is clearly visible that most of the residual values before the defect onset are close to zero. The outliers can be explained with scene properties (e.g. edges). In contrast, the residuals after the onset (around index 1770) are much higher.

To construct the feature space, we consider d different defects and their residuals. In this context, let $X = X_1, \ldots, X_d$ be the set of d features (Random Variables (RVs)) used for classification. With regard to the dynamic range of a pixel from 0 to 255, a single feature X_i can have a value $x_i \in \mathbb{Z} : -255 \leq x_i \leq 255$. If d features are regarded, then 511^d feature combinations fully define the feature space Ω. A single feature combination is denoted by the vector $\boldsymbol{x} \in [-255, 255]^d$, and it represents the observed residual values of a single image. The vector $\boldsymbol{x}$ can also be interpreted as point in Ω.

3.2   Naive Bayes Classifier (NB)

The NB (described in [1]) is a probabilistic classifier based on the Bayes Theorem, where

$$\hat{y} = \operatorname*{argmax}_{y \in Y} \; P(y|\boldsymbol{x}) = \operatorname*{argmax}_{y \in Y} \; P(y) \prod_{i=1}^{d} P(\boldsymbol{x}_{(i)}|y) \tag{3}$$

and Y denotes the set of k classes. To deduce $P(y|\boldsymbol{x})$ the conditional probabilities $P(\boldsymbol{x}_{(i)}|y)$ have to be estimated. In order to estimate $P(\boldsymbol{x}_{(i)}|y)$, the conditional probability distribution $p(X_i|\Psi)$ for defect X_i depending the defect's presence (i.e. $\Psi = \psi_1$ if the defect is present and $\Psi = \psi_0$ otherwise) has to be estimated. Depending on the class y and the defect X_i either $p(X_i|\psi_0)$ or $p(X_i|\psi_1)$ is considered. Hence, the estimation goal is to create k sets (one for each class), where each set contains d estimated conditional probability distributions, e.g. the first set consist of $\{p(X_1|\psi_0), \ldots, p(X_d|\psi_0)\}$ since there is no defect present.

Since X_i is discrete, an intuitive way to estimate $p(X_i|\Psi)$ is by the histogram of relative residual frequencies. However, the main problem with using a histogram is the likelihood of zero probabilities for certain bins. A way to prevent zero probabilities is through applying 'additive smoothing' [4], where a variable γ is added to every bin. This process smooths out the histogram by adding a uniform distribution. In order to estimate $p(X_i|\Psi)$, let $S_{X_i}^{(\Psi)} \subset S$ be a subset that contains observations of residual values of the ith defect, dependent on Ψ (before or after its onset).

$$\hat{p}(X_i|\Psi) = \frac{\gamma + \sum_{x \in S_{X_i}^{(\Psi)}} I(x_i = x)}{|S_{X_i}^{(\Psi)}| + \omega\gamma}, \tag{4}$$

where $\omega = 511$ and $I(.)$ is the indicator function returning 1 if the condition is true. A very small γ only adds a small probability to zero bins, while the original distribution is relatively unaffected. By increasing γ, more and more peaks are smoothed until only a uniform distribution remains (if $\gamma \to \infty$).

Another approach is to assume that $p(X_i|\Psi)$ follows a well defined probability distribution, i.e. that it is normally distributed. In this context, the estimation task is reduced to the estimation of the unknown parameters $\theta = (\mu, \sigma)$. This is done by computing the well known point estimators denoted as the empirical mean and standard deviation. The resulting probability distribution is continuous. Since X_i is a discrete RV, the probability of $P(x_i|\Psi) \; \forall \; x_i \in X_i$ given Ψ is

$$\hat{P}(z - \frac{1}{2} \leq z = x_i \leq z + \frac{1}{2}|\Psi) = \frac{1}{\sqrt{2\pi}\hat{\sigma}} \int_{z-\frac{1}{2}}^{z+\frac{1}{2}} \exp^{-\frac{(z-\hat{\mu})^2}{2\hat{\sigma}^2}} dz, \tag{5}$$

where $z \in \mathbb{R}$. However, the assumption of a normal distribution determines the distribution shape, regardless of whether this shape fits the actual residual distribution.

In contrast, non-parametric density estimation techniques provide more flexibility in the shape of the distribution. For this reason, we consider a Kernel Density Estimation (KDE) technique (described in [13]) as last approach. Again, the resulting distribution is continuous and

$$\hat{P}(z - \frac{1}{2} \leq z = x_i \leq z + \frac{1}{2}|\Psi)) = \frac{1}{|S_{X_i}^{(\Psi)}|h} \sum_{x \in S_{X_i}^{(\Psi)}} K\left(\frac{z-x}{h}\right). \tag{6}$$

A Gaussian kernel $K(.)$ is used, and h denotes the bandwidth.

The prior probabilities $P(y)$ of Eq. (3) can be estimated from the training set using the relative amount of images per class.

3.3   Support Vector Machine (SVM)

A SVM (described in [3]) is a geometric classifier, where the feature space is interpreted as d dimensional Cartesian space. In particular, the feature space Ω forms a d dimensional hypercube with an edge length of 511 (i.e. $[-256, 256]$). The goal of a SVM is to find the best separating hyperplane which divides Ω into two subspace, one for each class. For this reason, a SVM is a binary linear classifier predicting the positive class if

$$h(x) = \text{sgn}(\omega^T x + b) \tag{7}$$

results in a positive number. The separating hyperplane is defined by $\omega^T x + b = 0$. In order to apply a SVM to a multi-class classification problem, the problem has to be divided into several binary classification tasks. In particular, $\frac{k(k-1)}{2}$ SVMs must be trained in a one vs. one scenario. The resulting classifier is of the form

$$\hat{y} = \underset{y \in Y}{\text{argmax}} \left(\sum_{h_y(x) \in H_y} h_y(x) \right), \tag{8}$$

where H_y is a set containing the hyperplane functions h_y between class y and all other classes.

4   Practical Implementation Issues

To apply the proposed classifiers, the defect locations have to be known beforehand. In [8], the authors suggest to threshold the residual variance in order to define the defect candidates. Other in-field sensor defect detection techniques are introduced in [6] and [14]. Besides the defect locations, the onset times have to be known as well. To estimate the onset time, we exploit the difference in the average residual value before and after the defect onset (i.e. we estimate j where the difference is maximum). This approach works well for most defects. However, since the onset times define the class borders (i.e. the classes), knowing the correct onset times is crucial for the subsequent image age approximation. For this purpose, a visual refinement of the onset times is necessary, i.e. the estimated onset times have to be compared to the onset times defined by looking at the chronologically ordered residuals (e.g. as shown in Fig. 2).

Furthermore, it is possible that multiple defects have their onset times at the same index j. One explanation for this could be a higher exposure to cosmic radiation (e.g. caused by transport on an airplane) at this point in time. Alternatively, there could be a significant time difference between the real onset times; however, no images were available during this time. To deal with this possibility, similar onset times are clustered using hierarchical clustering with a single linkage and a cut-off depth $t = 20$. As a result, there are at least 20 images between all onset times.

To process color images, the authors in [8] convert the images into grayscale images beforehand. Since a defect is not always represented the same way in each color channel, a grayscale conversion can attenuate the defect intensity. To cope with this, we select the color channel for each defect in which the average residual value (over all training data) is maximum.

The estimation of $p(X_i|\Psi)$ via the histogram with additive smoothing can bias the age prediction result. This occurs when the relative residual frequency of a certain value is equal before and after the defect onset. Because of the additive smoothing property, the resulting probability is higher when fewer samples are used for computation. For example, if the occurrence of 255 is 0 by taking residual values into account either before or after the onset, and if fewer samples are available before the onset, then the resulting probability of 255 before the defect onset is higher.

A general problem for both continuous density estimation techniques are the resulting probability values for marginal areas near -255 and 255 since the probabilities are very low (probably below machine precision) at these points and are set to zero.

5   Experiments

The age prediction accuracy of the classifiers is empirically evaluated based on several experiments. For a better comparison between the proposed classifiers and the maximum likelihood approach introduced in [8], we implemented this method as a classifier and applied it to the same data sets. In this context, the maximum likelihood approach of Eq. (2) changes to

$$\hat{y} = \underset{y \in Y}{\arg\max} \prod_{i \in \Omega} \frac{1}{\sqrt{2\pi}\hat{\sigma}_{(i)}^{(y)}} \exp \frac{W_{(i)} - \left(I_{(i)}K_{(i)}^{(y)} + \tau D_{(i)}^{(y)} + c_{(i)}^{(y)}\right)}{2\hat{\sigma}_{(i)}^{2(y)}}. \tag{9}$$

The defect parameters $(K^{(y)}, D^{(y)}, c^{(y)}, \sigma^{(y)})$ are computed according to the onset times defined by class y. All the experiments rely on images from three different imagers, a Nikon E7600, a Canon PowerShot A720IS and a Sony DSC-P8.

5.1   Data Sets

Let S_N be a set of 1768 chronologically ordered images that were captured with the Nikon. The set S_C consists of 4379 chronologically ordered images taken with the Canon and 2302 images captured with the Sony are in set S_S. Figure 3 shows the relative amount of images taken per year. The images of all three imagers were taken between 2003 and 2014. All devices were used as personal cameras to capture regular scene images (e.g. vacation scenes). A sample of the captured scenes from all three devices is illustrated in Fig. 4, in which 4 images from each device are shown. The images captured in the years 2019 and 2020 were taken for this paper. These recently taken images were only captured with the Nikon and Canon as only those devices are still available.

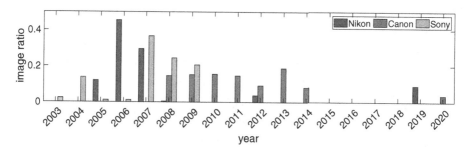

Fig. 3. Relative amount of images per year and device.

A 7 megapixel Charged Coupled Device (CCD) sensor with a resolution of 3072×2304 is used for the Nikon, and a 8 megapixel CCD sensor with a 3264×2448 resolution is built into the Canon. The native Canon sensor resolution is altered by a quality setting, whereby the images are stored in a format of 2592×1944. All Sony images were taken with a 3.2 megapixel CCD sensor with a resolution of 2048×1536. All images are JPEG compressed 8 bit RGB color files. On the basis of the given JPEG compressed datasets, an evaluation of different compression ratios would imply 'double compression', which is of less relevance.

Fig. 4. Randomly sampled images from set S_N, S_C and S_S.

Based on the onset time analysis, 8 classes were identified for the Nikon and 7 each for the Canon and Sony. In Fig. 5, the relative ratio of images and defect onsets per class is illustrated. The first class always represents the time interval between the first available image and the first defect onset. For this reason, there is no defect present in the first class. In total, 23 defects spreading over 87 pixels are found on the Nikon sensor. For the Canon, 17 defects spreading over 65 pixels are used for age prediction, and 8 defects spreading over 42 pixels are used for the Sony. The large proportion of defects in the last Canon and Nikon class is due to the huge time gap between the last and penultimate class (as illustrated in Fig. 3).

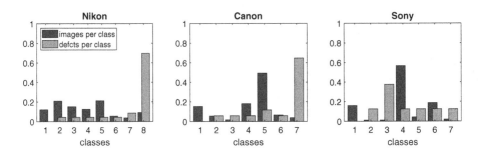

Fig. 5. Relative amount of images and defects per class.

5.2   Performance Metric

The accuracy of a classifier can be expressed in terms of the probability of an arbitrary object being misclassified. This error is known as the generalization error. Unfortunately, it is not possible to compute this probability as only a finite number of samples are available. However, it is possible to compute the empirical error

$$L_S(\hat{y}) = \frac{1}{|S|} \sum_{i=1}^{n} I[\hat{y} \neq y], \tag{10}$$

where I is the indicator function returning 1 only if the argument is true. The empirical error is computed from a data set, also known as test set, that the classifier has not seen before. In this context, the empirical error is an approximation of the generalization error. To compute the empirical error, we divide the available data sets into a training and test set. This is done in a stratified manner with a sampling ratio of 0.9. Since a stochastic process is used for sampling, more than one training and evaluation operation is conducted. In particular, 200 different evaluations are carried out.

A natural upper classification bound is the empirical error that results from predicting by chance (i.e. sampling from a uniform distribution $U_{1/|Y|}$). Consequently, the resulting upper bound of the empirical error is equal to $1 - 1/|Y|$.

In order to evaluate the prediction performance for a single class, we computed the f1-score

$$\text{f1} = \frac{2\text{TP}}{2\text{TP} + \text{FP} + \text{FN}} \tag{11}$$

for each class. To compute the f1-score for a single class, this class is regarded as positive class, and all other classes are denoted as negative class. Taking into account a random prediction (for all classes), the lower f1-score bound of the ith class is

$$\frac{2\frac{\alpha N}{|Y|}}{2\frac{\alpha N}{|Y|} + \frac{(1-\alpha)N}{|Y|} + \alpha N - \frac{\alpha N}{|Y|}} = \frac{2\frac{\alpha N}{|Y|}}{\frac{N + |Y|\alpha N}{|Y|}} = \frac{2\alpha}{1 + \alpha|Y|}, \tag{12}$$

where N is the sample size of the test set and $n_i = \alpha N$ are the number of samples from the ith (positive) class. The amount of samples from the negative class (all other classes) is represented by $(1 - \alpha)N$. If the test data are uniformly distributed over all classes, then $\alpha = 1/|Y|$ and the lower f1-score bound is equal to $1/|Y|$ (as expected). If we assume a uniform random prediction for a single class only, the lower f1-score bound is

$$\frac{\frac{2\alpha N}{|Y|}}{\frac{2\alpha N}{|Y|} + \alpha N - \frac{\alpha N}{|Y|}} = \frac{\frac{2\alpha N}{|Y|}}{\frac{\alpha N + |Y|\alpha N}{|Y|}} = \frac{2}{1 + |Y|}. \tag{13}$$

This results from assuming $FP = 0$, i.e. no samples from other classes are predicted as this class (precision $= 1$). Hence, if any method yields a higher f1-score than $2/(1 + |Y|)$ for a single class, the prediction result for this class is definitely better than a random prediction.

6   Results

Table 1. Represents the average empirical error (over 200 runs). The left value shows the empirical error that results when it is assumed that the class probabilities follow the training data distribution, and the right value results when uniform class probabilities are assumed.

Device	KDc	NB-NE	NB-HE	NB-KDE	SVM
Nikon	0.46	0.46/0.47	0.35/0.37	0.38/0.39	0.38/0.38
Canon	0.28	0.27/0.31	0.20/0.24	0.21/0.26	0.19/0.22
Sony	0.33	0.28/0.34	0.25/0.36	0.21/0.33	0.21/0.30

In Table 1 the empirical error for each method is shown. The 'KDc' method represents the results of applying the technique introduced in [8]. This method is used as reference as it represents the current state-of-the-art in this field. All NB results are denoted by 'NB-NE', 'NB-HE' and 'NB-KDE', where $p(X_i|\Psi)$ is estimated either by assuming a normal distribution ('NB-NE'), via a histogram ('NB-HE'), or through the KDE ('NB-KDE'). Both parameters of the histogram and the KDE are treated as hyperparameter, with the smoothing parameter $\gamma = 0.028$ and the bandwidth h is set to 2.4 (for the Nikon and Canon) and 4.2 for the Sony.

The left values in Table 1 represent the empirical error when applying the classifiers as stated in Sect. 3 (i.e. NB with priors $P(y)$ according to the relative amount of training data per class and the standard matlab implementation of the SVM). Overall, the SVM achieves the best error rates, which are significantly lower compared to the current state-of-the-art ('KDc'). Compared to the SVM, the error rates resulting from the NB using a histogram and a KDE probability distribution estimation technique are only slightly worse but still significantly lower than those produced using the 'KDc' method. However, these large differences are based on the assumption that the image class distribution of the untrustworthy set follows the distribution of the training data (trusted set), and this assumption is probably not applicable to real world data. For this purpose, we have additionally trained the NB with uniform priors. Since the training data are not uniformly distributed across the classes, the SVM tends to overfit classes with more training data. In order to compensate this overfitting, we also impose a uniform class distribution when training the SVM. This problem of unbalanced classes is less important for learning the probability distributions $p(x|\Psi)$ since all data, either before or after the defect onset, are always taken into account (i.e. across the class boundaries).

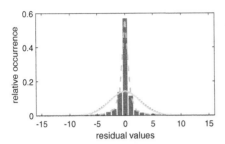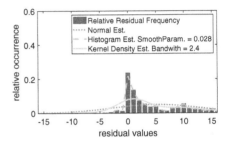

Fig. 6. Illustrates the different probability distribution estimation techniques. The left figure shows the relative residual frequencies before the defect onset and the right figure illustrates the relative frequencies afterwards.

The empirical error rates when assuming a uniform class distribution are represented by the right values in Table 1. The SVM again achieves the best results, which are considerably better than the 'KDc' method. In total, the second best results are achieved with the 'NB-HE'. Nevertheless, the 'NB-KDE' achieves almost the same low error rates. As the results indicate, assuming a normal distribution of residual values is less correct. This claim is confirmed by looking at Fig. 6. The relative residual frequency (before and after the onset) of a certain defect is compared with the different density estimation techniques. It is clearly visible that a normal distribution fits the residual distribution relatively well before the onset, but not at all afterwards. The distribution shapes illustrated look quite reasonable since regular scenes usually consist of many smooth areas, which are represented by the high probability that a residual before the onset is zero (see Fig. 6 (left)). The probabilities directly to the left and right of zero are due to scene properties such as borders between two smooth areas (edges). A negative residual value occurs when there are more 'dark' than 'bright' pixels in the 3×3 median filter area, and the middle pixel is a bright one. A contrary situation would cause a positive residual value. The symmetry of probabilities is reasonable since these described situations are likely to be equally probable. After the defect onset (see Fig. 6 (right)), a negative residual value is less likely. This is due to the defect properties, where the defective pixel value is usually higher than its neighbors.

A residual value of zero (or near to zero) after the defect onset can be caused by bright scenes or by the processing pipeline and image compression (i.e. attenuation of high frequencies). This is a problem when predicting the correct image age since a residual of zero always indicates a not present defect. On the basis of a feature analysis, there is an estimated probability of 0.2113, 0.1330 and 0.1847 (Nikon, Canon and Sony) that a residual value after the defect onset is zero. These probabilities reflect the difference in the processing pipeline and image compression. Which most likely causes the differences observed between all three devices.

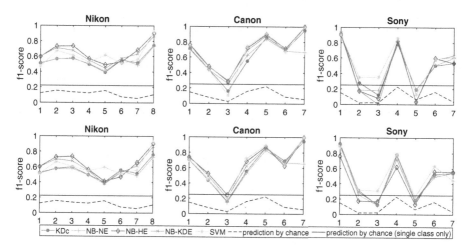

Fig. 7. Represents the average (over 200 runs) f1-score for each class. The top row represents the f1-score when the class probabilities are assumed to follow the training data distribution, and the bottom row represents the f1-score when uniform class probabilities are assumed.

To asses the prediction performance of the individual classes, we evaluated the f1-score for each class. In Fig. 7, the average f1-score (over all 200 runs) per class is illustrated. The top row shows the f1-score per class and device, assuming that the class probabilities follow the training data distribution, and the bottom row represents the f1-score when uniform class probabilities are assumed. The solid black horizontal line represents the derived lower bound of Eq. (13). A f1-score higher than this line indicates a prediction result definitely better than a random selection. The dashed black line reflects the weaker lower bound, where a general random prediction (for every class) is assumed. This lower bound is derived from Eq. (12) and is dependent on the sample class ratio.

The Nikon is the only device where all methods are definitely better than random predictions (given all classes). With the Canon, the f1-score of class 3 yields the worst result. This might be due to the imbalance in the amount of images between the classes. The Canon class 3 has the lowest amount of images as compared to the other classes. The same argument applies to the Sony, where the classes 2, 3 and 5 yield the lowest f1-scores. A positive generalization effect when assuming uniform class probabilities (for NB and SVM) is clearly visible, i.e. the prediction performance is higher for classes with fewer images. The very high f1-score of the last class for the Nikon and Canon is due to the particularly high amount of defects that occur in this class (due to the huge time difference).

7   Conclusion

Approximating the age of a digital image based on the presence of in-field sensor defects is a multi-class classification problem where a class is defined as the time interval between two consecutive defect onset times. For this purpose, the temporal prediction accuracy depends on the number of different defect onsets in the available time interval. To solve this classification problem, standard machine learning techniques are considered. In particular, a probabilistic (NB) and a geometric classifier (SVM). In general, the SVM achieves the best age prediction results, which are considerably better than the current stat-of-the-art introduced in [8]. Thus, the described SVM is the recommended technique to approximate the age of an image under investigation. Overall, the second best results are achieved by the 'NB-HE'. However, due to the small smoothing parameter γ, this method tends to overfit the data. As illustrated in Fig. 6, the kernel density estimation generalizes the observed relative residual frequency to a higher extent. For this reason, this density estimation technique is preferable over all other estimation techniques described.

In general, the more images available for training and the more defects per class, the better the results. A higher amount of defects is particularly helpful in predicting the age of an unseen image since the probability is higher that not all defects are attenuated due to local scene properties (e.g. bright areas), or due the differences in processing pipelines and image compression. We remind that the observed probability of a residual value being zero after the defect onset is 0.2113, 0.1330 and 0.1847 (Nikon, Canon and Sony), which indicates an expected significant prediction error.

In conclusion, image age approximation based on in-field sensor defects is possible. In particular, when a class is defined by multiple defects, the accuracy for that class is very high, i.e. an f1-score of 0.9976 for the Canon class 7 and 0.9227 for the last Nikon class.

References

1. Barber, D.: Bayesian Reasoning and Machine Learning. Cambridge University Press, Cambridge (2012). https://doi.org/10.1017/cbo9780511804779
2. Chapman, G.H., Thomas, R., Koren, Z., Koren, I.: Empirical formula for rates of hot pixel defects based on pixel size, sensor area, and ISO. In: Widenhorn, R., Dupret, A. (eds.) Sensors, Cameras, and Systems for Industrial and Scientific Applications XIV, vol. 8659, pp. 119–129. International Society for Optics and Photonics, SPIE (2013). https://doi.org/10.1117/12.2005850

3. Chen, L.-P., Mehryar, M., Afshin, R., Ameet, T.: Foundations of machine learning. Stat. Pap. **60**(5), 1793–1795 (2019). https://doi.org/10.1007/s00362-019-01124-9. Second edition

4. Chen, S.F., Goodman, J.: An empirical study of smoothing techniques for language modeling. Comput. Speech Lang. **13**(4), 359–394 (1999). https://doi.org/10.1006/csla.1999.0128

5. Dudas, J., Wu, L.M., Jung, C., Chapman, G.H., Koren, Z., Koren, I.: Identification of in-field defect development in digital image sensors. In: Martin, R.A., DiCarlo, J.M., Sampat, N. (eds.) Digital Photography III, vol. 6502, pp. 319–330. International Society for Optics and Photonics, SPIE (2007). https://doi.org/10.1117/12.704563

6. El-Yamany, N.: Robust defect pixel detection and correction for Bayer imaging systems. Electron. Imaging **2017**(15), 46–51 (2017). https://doi.org/10.2352/issn.2470-1173.2017.15.dpmi-088

7. Fridrich, J.: Sensor defects in digital image forensic. In: Sencar, H., Memon, N. (eds.) Digital Image Forensics. Springer, New York (2013). https://doi.org/10.1007/978-1-4614-0757-7_6

8. Fridrich, J., Goljan, M.: Determining approximate age of digital images using sensor defects. In: Memon, N.D., Dittmann, J., Alattar, A.M., III, E.J.D. (eds.) Media Watermarking, Security, and Forensics III, vol. 7880, pp. 49–59. International Society for Optics and Photonics, SPIE (2011). https://doi.org/10.1117/12.872198

9. Leung, J., Chapman, G.H., Koren, I., Koren, Z.: Characterization of gain enhanced in-field defects in digital imagers. In: 2009 24th IEEE International Symposium on Defect and Fault Tolerance in VLSI Systems, pp. 155–163 (2009). https://doi.org/10.1109/DFT.2009.49

10. Leung, J., Chapman, G.H., Koren, Z., Koren, I.: Statistical identification and analysis of defect development in digital imagers. In: Rodricks, B.G., Süsstrunk, S.E. (eds.) Digital Photography V, vol. 7250, pp. 272–283. International Society for Optics and Photonics, SPIE (2009). https://doi.org/10.1117/12.806109

11. Leung, J., Dudas, J., Chapman, G.H., Koren, I., Koren, Z.: Quantitative analysis of in-field defects in image sensor arrays. In: 22nd IEEE International Symposium on Defect and Fault-Tolerance in VLSI Systems (DFT 2007), pp. 526–534. IEEE (September 2007). https://doi.org/10.1109/dft.2007.59

12. Piva, A.: An overview on image forensics. ISRN Signal Process. **2013**, 1–22 (2013). https://doi.org/10.1155/2013/496701

13. Silverman, B.W.: Density Estimation for Statistics and Data Analysis, vol. 26. CRC Press, Boca Raton (1986)

14. Tchendjou, G.T., Simeu, E.: Detection, location and concealment of defective pixels in image sensors. IEEE Trans. Emerg. Top. Comput. 1 (2020). https://doi.org/10.1109/tetc.2020.2976807

15. Theuwissen, A.J.: Influence of terrestrial cosmic rays on the reliability of CCD image sensors part 1: experiments at room temperature. IEEE Trans. Electron. Devices **54**(12), 3260–3266 (2007). https://doi.org/10.1109/TED.2007.908906

16. Theuwissen, A.J.: Influence of terrestrial cosmic rays on the reliability of CCD image sensors part 2: experiments at elevated temperature. IEEE Trans. Electron. Devices **55**(9), 2324–2328 (2008). https://doi.org/10.1109/TED.2008.927662

Facing Image Source Attribution on iPhone X

Daniele Baracchi[1], Massimo Iuliani[1,2], Andrea G. Nencini[1],
and Alessandro Piva[1,2(✉)]

[1] Department of Information Engineering, University of Florence, Florence, Italy
alessandro.piva@unifi.it
[2] FORLAB, Multimedia Forensics Laboratory, PIN scrl, Prato, Italy

Abstract. Most image forensics techniques rely on the analysis of traces left into the signal during the image acquisition process, which is supposed to be common among most devices. However, recent advances in visual technologies led several manufacturers to customize the acquisition pipeline in order to improve the image quality, by designing alternative coding schemes and in-camera processing. This fact threatens the effectiveness of available forensic techniques. It is thus required to study modern acquisition devices to both assess the effectiveness of available techniques and to develop new effective approaches.

In this paper we focus on the source identification of images coming from one of the most spread moderm smartphones, i.e. the iPhone X. This model is significant since it comprises two main new features: the new HEIF compression standard, set as the default image format, and a brand new shooting mode, called *Portrait Mode*. We show that existing source identification methods are ineffective when images are acquired in *Portrait Mode*. We also show when and how it is possible to address this limitation by removing non unique artefacts introduced by the camera software.

Keywords: Image forensics · Source identification · PRNU · Smartphones

1 Introduction

Image authentication is based on the ability to verify the presence and consistency of traces left into the image by the acquisition pipeline. Such a process is usually represented as a composition of several steps [18]: the light is focused by the lenses on the camera sensor and generally filtered by a CFA (Color Filter Array), that selectively permits a certain component of light to pass through it to the sensor. The three chromatic components of each pixel are then rebuilt by a demosaicing interpolation. Additional in-camera processing (e.g. white balancing, gamma correction) are then applied to the image before the final compression, usually in JPEG format, and storage to the camera memory.

Most forensic technologies assume the above image life cycle. For instance, one well-known tampering detection technique is based on the hypothesis that

© Springer Nature Switzerland AG 2021
X. Zhao et al. (Eds.): IWDW 2020, LNCS 12617, pp. 196–207, 2021.
https://doi.org/10.1007/978-3-030-69449-4_15

a tampered image underwent a double JPEG compression [1], one during the acquisition and another one after the tampering operation. However, this hypothesis certainly will not stand long due to the introduction of the new HEIF image compression format on some recent smartphone models. Another relevant example is related to the source identification based on the unique traces left into the image or video by to the Photo Response Non Uniformity (PRNU) [3,16]. This trace allows to assess whether a picture belongs to a specific device by comparing noise pattern extracted from images. The effectiveness of this technique is based on two main assumptions: (i) the reference pattern and the compared images are pixel-by-pixel aligned and (ii) in-camera processing do not compromise the noise pattern left during the acquisition process. Researchers already highlighted that the first hypothesis does not stand anymore on most devices since Electronic Image Stabilization (EIS) was introduced [11] into the video acquisition pipeline. On the contrary, the second issue has been neglected until now since the image generation process of most digital cameras does not ruin the sensor pattern noise. However, new introduced coding schemes (e.g., HEVC in HEIF images) and shooting modes (e.g., bokeh mode, night sight mode) strongly affect the outputted signal thus posing in serious danger the applicability of the technique.

In this paper, we begin addressing these issues by analyzing the Apple iPhone X since, starting from this model, Apple set the default image format to HEIF. Furthermore, this device allows to shot images in *Portrait Mode*, which causes the background and the foreground to be processed in very different ways. This model also represents a relevant reference being one of the most popular devices (the best selling smartphone in 2018 [8]). Firstly, we investigate the presence of PRNU traces on HEIC images in comparison to their JPEG counterparts. Then, moving to the brand new *Portrait Mode* we highlight that in-camera processing introduces a correlation artifact in the background that leads to extremely high correlations even with images belonging to different devices. In this regard, we show when and how it is possible to address this limitation in practical scenarios by removing those non-unique artifacts.

The paper is organized as follow: in Sect. 2 a brief description of HEVC coding and an introduction to the *Portrait Mode* based on available patents are given; in Sect. 3, we summarize how PRNU-based source identification works and we define how to address the issues raised in the *Portrait Mode*; in Sect. 4 we describe the performed experiments and we report the achieved results.

2   HEIC Format and Portrait Mode

HEIF (ISO/IEC 23008-12) [10] is a container format based on *ISO Base Media File Format* [9], capable of encapsulating individual images or sequences. HEIF standard supports different compression algorithms for the payload [7], but typically uses HEVC: in this case, the format is also known as HEIC. More specifically, although the HEVC/H.265 codec is designed for video streams, the HEVC intra-frame encoding is exploited by HEIF specifications to compress single images. Similarly to JPEG, HEIC images are split into blocks that are processed in the frequency domain by means of quantization and entropy encoding of transformed coefficients. However, the HEVC algorithm provides, under comparable compression factors, higher quality images: this is done by using variable-size blocks (64×64, 32×32, 16×16, 8×8), by implementing adaptive deblocking filters and *Sample adaptive offset* (SOA) filtering for a better reconstruction of the original values, and by encoding the quantized coefficients using *context-adaptive binary arithmetic coding* (CABAC).

With regard to the acquisition phase, the iPhone X is equipped with two rear cameras: an f/1.8 wide lens camera and an f/2.4 telephoto sensor. When default settings are used, images are taken with the wide lens camera; when *Portrait Mode* is activated, the phone switches to the telephoto sensor[1]. This fact by itself highlights that, in the source identification task, these images cannot be generally compared with images captured with the phone in default settings since they are acquired using a completely different sensor. Given a main subject in the foreground (e.g., a face), *Portrait Mode* outputs an image with a blurred background and a possibly enhanced subject (example in Fig. 2). This implies that the image is further processed in-camera.

Implementation details of this shooting mode are not publicly available; however, the macro-structure of the *Portrait Mode* pipeline is described by some Apple's patents [2, 15]. The system estimates the depth of scene points by stereo pair triangulation and creates a stack of progressively blurred versions of the acquired image, one for each desired plane of depth. Then the final image is crafted by selecting the pixels of a plane of depth from the appropriate blurred image in the stack. During the acquisition, the depth is likely produced with the help of the wide lens camera; this is confirmed by the fact that the obstruction of the f/1.8 lens with a finger prevents the depth map creation while shooting in *Portrait Mode*.

Eventually, this acquisition produces three files:

1. A *base portrait* image (see Fig. 1) exported as HEIC or JPEG, usually according to the option set by the user. The image resolution is the same of *standard images* (4032×3024). The image file also contains the depthmap of the scene as a binary EXIF metadata. The grayscale depthmap (see Fig. 3), with size 768×576, records the distances of scene points from the camera. Based on the conducted experiments, the depthmap is created by the camera software only if the scene is not completely flat.
2. An AAE file, representing a *property list* [19], an XML file that stores objects and settings of Apple applications. This particular file defines a dictionary of obscure key-value pairs for the iOS *Photos* application[2].
3. A *bokeh portrait* image with blurred background (see Fig. 2), always exported in JPEG format, generated with a post-processing pipeline that involves the

---

[1] Information present in the image metadata. We noticed that the telephoto sensor is exploited also when the zoom is activated in standard mode.
[2] To the best of our knowledge there is no public documentation of this file's content.

original image, the depthmap and maybe the AAE file. *Bokeh portrait* images also store the depthmap of the scene, like base images, in their EXIF metadata.

In the end, the *bokeh portrait* image is usually the one chosen by the user since it appears as an aesthetically enhanced image where the background is blurred to highlight the represented subject (a face or an object).

Fig. 1. *Base portrait* image.   **Fig. 2.** *Bokeh portrait* image.   **Fig. 3.** The scene depthmap.

It should be noted that the above considerations stand for both iPhone X and iPhone XS. However, in our tests we also considered an iPhone XR that presents an important technological difference: it is equipped with a single rear camera (an IR sensor is exploited to compute image depth), so that *Portrait* and default photos are always acquired with the same sensor. This fact is extremely important to highlight the main idea inspiring this paper, i.e. that the acquisition pipeline and in camera processing are becoming model dependent, thus requiring forensic analyses on specific brand and models.

3 Source Identification

In this section we summarise the PRNU-based image source identification and the method adopted to mitigate the false alarm introduced by the *Portrait Mode*.

3.1 Image Source Identification

The image source identification based on PRNU is organized in two steps: i) a reference sensor fingerprint is derived from still images acquired by the source device; ii) given a query image, its noise residual is estimated and then compared with the reference fingerprint to assess a possible match.

The camera fingerprint **K** can be estimated from N images $\mathbf{I}_1, \ldots, \mathbf{I}_N$ captured by the source device. A denoising filter [13,14] is applied to each image and the noise residuals $\mathbf{W}_1, \ldots, \mathbf{W}_N$ are obtained as the difference between each image and its denoised version. Then, the fingerprint estimate $\widetilde{\mathbf{K}}$ is derived by the maximum likelihood estimator [4]:

$$\widetilde{\mathbf{K}} = \frac{\sum_{i=1}^{N} \mathbf{W}_i \mathbf{I}_i}{\sum_{i=1}^{N} (\mathbf{I}_i)^2}. \tag{1}$$

Given the probe image **I**, its noise residual **W** is estimated and the test statistic is built as the peak to correlation energy (PCE) between $\mathbf{I}\widehat{\mathbf{K}}$ and **W** [6]. It is expected that images belonging to the reference device will exhibit a PCE well higher than a certain threshold (usually accepted around 60 [6]), whereas images taken by other devices, also of the same model, will show low values of PCE.

3.2 Image Source Identification in *Portrait Mode*

As we will show in the experimental section, such an approach is proved to work on images taken in default settings. Unfortunately, strong correlation peaks also appear on images belonging to different devices when captured in *Portrait Mode*. To highlight this fact, in Fig. 4 we compare the local PCE between two *base portrait* photos of different devices and their corresponding *bokeh portrait* versions[3]. The first case results in a PCE peak of 26, as expected lower than the threshold; however, in their processed versions a huge local PCE peak of 47401 appears due to the image post processing.

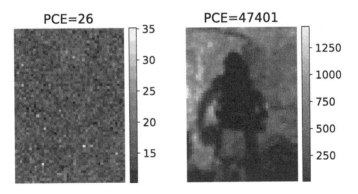

Fig. 4. Local PCE between two *base portrait* photos from different devices (Left) and between the *bokeh portrait* version of the same two images (Right).

We found this behavior in all bokeh portrait photos of our datasets and we attribute it to the presence of some non-unique artifact that is concentrated on

---

[3] Local PCE is computed by a 128×128 sliding window with 64 pixel shift.

blurred regions, as highlighted by the brighter areas in Fig. 4. The existence of this artifact is insidious for source identification, because its presence can lead to mismatching attributions.

In order to mitigate this critical issue we adopted two main strategies: first of all, in the PCE computation we limit the accepted peaks in a 10 pixel radius circle centred in the frame origin, since when the image is taken in *Portrait Mode* we do expect, at most, very subtle cropping operations to build the final enhanced image. On the contrary, we set the PCE to 0 when the peak is found elsewhere. This method allows to reduce false alarm rate. Furthermore, given that the background post processing introduces a strong correlation artifact, we exploited the depth map information to remove such post processed image parts, as detailed in the following.

Pre-processing of a Probe *bokeh portrait* Image. By assuming that blurred areas give little contribution to the height of the correlation peak (because of the attenuated PRNU), the idea is to replace those areas with zeros to decrease the energy of the signal. This is accomplished by means of the binarization of the depth map to remove the (blurred) background. More specifically, given the tested grey scale image $\mathbf{I}$, its depth map $\mathbf{D}$ is converted into a binary matrix $\bar{\mathbf{D}}$ by choosing the threshold that maximizes the separability of foreground and background [17]. Then, the image $\mathbf{I}$ is pixel wise multiplied by $\bar{\mathbf{D}}$ to remove the background. This processed image will then be ued for the computation of PCE.

Pre-processing of a Reference Fingerprint Computed from *bokeh portrait* Image. If the reference fingerprint has to be estimated from *bokeh portrait* images, it is not possible to exploit in a straightforward way the previous method, since the blurred background is present in different areas of each image. To cope with this, we considered two ways to remove the traces of non unique artifacts:

In this scenario we tried to reduce artefact traces as much as possible also in the reference signal, and we designed two strategies that operate on the estimated fingerprint $\mathbf{K}$:

- The first method (indicated as *weighted fingerprint*) requires, when estimating the fingerprint from N images, to count for each pixel in how many images that pixel belongs to the foreground region. At the end the fingerprint $\mathbf{K}$ is multiplied with the normalized occurrence matrix to obtain the weighted fingerprint $\mathbf{K_w}$:

$$\mathbf{K_w} = \mathbf{K} \cdot \frac{\sum_{i=1}^{N} \bar{\mathbf{D}}^{(i)}}{N} \tag{2}$$

with $\bar{\mathbf{D}} \in \{0,1\}^{m \times n}$ being the binarized depth map, in which a value of 1 indicates a foreground pixel and 0 a background pixel.

In this way, fingerprint regions estimated from foreground areas will be more relevant in correlation, and, at the same time, the most artifact-affected regions are not completely discarded, assuming that they preserve some traces of PRNU.

– The second method (indicated as *binary fingerprint*) is more drastic in excluding artifacts from the fingerprint; background areas of every image involved in the estimation are removed from the fingerprint by replacing them with zeros, as in probe signal background exclusion[4]. The fingerprint $\mathbf{K_b}$ is computed as:

$$\mathbf{K_b} = \mathbf{K} \cdot \prod_{i=1}^{N} \bar{\mathbf{D}}^{(i)} \tag{3}$$

Note that $\mathbf{K_w}$ weighs pixels based on their frequency in the image foregrounds; $\mathbf{K_b}$ only considers the pixels that belong to the foreground in all images.

4   Experimental Results

In this section, we describe the results of the experiments performed to evaluate the performance of source identification, when working with HEIC images and then when dealing with Portrait Images.

4.1   Source Identification on HEIC vs. JPEG Images

The first experiment focuses on assessing the presence of PRNU trace in *HEIC* images with default settings[5].

We considered an iPhone X, iOS 12.1.4, and we estimated two sensor fingerprints from 50 flat images in *HEIC* and *JPEG* format respectively. For each fingerprint, we computed the PCE with 40 images of natural scenes captured in both *HEIC* and *JPEG* format by the device. In Fig. 5 we report the PCE histograms, where the label at the top refers to reference and test image formats, in this order. Although the tested image contents are not the same, there is statistical evidence that JPEG images expose higher correlations. This fact could be attributed both to a stronger presence of the PRNU trace and to non-unique artifacts left by the denoising process onto JPEG images. However, JPEG artifacts would only appear when both reference and test belong to JPEG format (fourth histogram). Here, we note that correlation grows in second and third histograms too, where the reference or the probe only belongs to JPEG, thus providing strong evidence that the peak difference is more likely attributable to the HEVC coding that undermines the PRNU traces. To better emphasize this

---

[4] Obviously, each background exclusion can erode the content of the fingerprint, and exists the risk of reducing the signal to an excessively small patch. To avoid this effect, it is recommended to estimate the fingerprint from photos that present large foreground areas localized in the same region of the image surface.

[5] Note that, due to the poor HEIC support of MATLAB and Python modules, HEIC images have been converted to PNG, a lossless format, with *libheif* [12]; this procedure maintains the signal unaltered and allows reading the images with several software libraries.

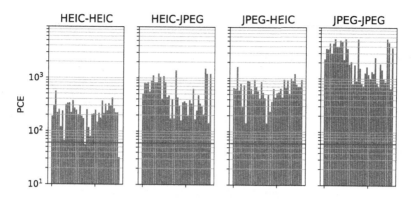

Fig. 5. PCE histograms, obtained comparing references in *HEIC* and *JPEG* format with both *HEIC* and *JPEG* residuals. The red horizontal line marks the threshold value. The label at the top refers to reference and test image formats, in this order. (Color figure online)

fact, for each considered test, we also computed 40 mismatching PCE values with images belonging to an iPhone 6.

Table 1. Accuracy and max threshold to achieve it.

Reference	Test	Accuracy	Threshold
HEIC	HEIC	0.975	55.7
HEIC	JPEG	1.0	143.7
JPEG	HEIC	1.0	143.9
JPEG	JPEG	1.0	649.6

In Table 1 we report the accuracies and the max threshold that can be used to achieve them. In HEIC-HEIC case, a threshold of 55.7 already causes a mismatching.

Note that, although a threshold of 60 is commonly accepted as a guarantee of low false alarm [5], JPEG images allow to set a much higher threshold (over 600) while still granting perfect classification. This does not hold for HEIC images, where the compression performed on images hinders the PRNU traces, thus leading to harder-to-detect correlations. Based on these results, we should expect lower robustness of the PRNU trace to compression and other processing since it is already deteriorated by the HEVC encoding.

4.2   Source Identification of Portrait Images

We considered then an iPhone X, iOS 12.1.4, whose fingerprint was computed from 50 images through Eq. 1. Tests were performed on 100 portrait images captured with the same device and 100 images taken with other similar devices (iPhone X, iPhone XR, and iPhone XS Max). We considered three different scenarios:

Case 1: the device is available and the tested image is a *base portrait* photo. Here, we estimate the fingerprint from *base portrait* images representing flat scenes allowing to compute a good sensor reference pattern estimate. Furthermore, tested images are possibly untouched by further in-camera post processing. In this case all the images are correctly classified (AUC = 1), as shown in Fig. 6.

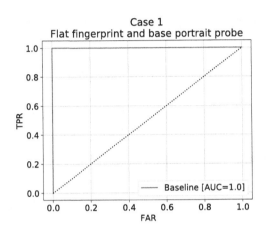

Fig. 6. Case 1: the fingerprint is estimated from flat images and the test image is a *base portrait*.

Case 2: the device is available and the tested image are *bokeh portrait* photos, i.e. they are internally processed resulting in blurred backgrounds. Here, we applied both the baseline method and the strategy proposed in Sect. 3 to remove the background with the Otsu thresholding and to limit the PCE in the strict neighbours of the image frame. The reference is estimated as in *Case 1*. In Fig. 7 we report the ROC curve using the baseline method and the proposed approach. Note that, with 0 FAR, the TPR increases from 0.73 to 0.87.

Case 3: the device is not available and both reference and test are bokeh portrait photos. In this case, the tested images are treated as in *Case 2*; furthermore we try to remove the non unique artifact on reference side too by computing the *weighted* and *binary* fingerprints (see Eqs. 2 and 3). In Fig. 8 the ROC curve is reported for baseline method and the proposed approaches. It can be clearly seen that the baseline method is ineffective in this scenario and that the proposed method employing the *binary* fingerprint provides a significant improvement (TPR at 0 FAR shifts from 0 of the baseline to 0.45, and reaches 0.64 at 5% FAR). Note that in the weighted approach some background pixels

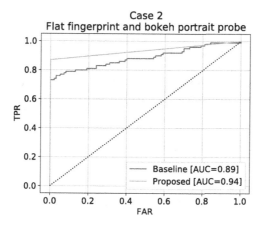

Fig. 7. Case 2: the fingerprint is estimated from flat images and the test is a *bokeh portrait* image.

contribute to the fingerprint formation. As the non-unique artifact in the background strongly correlates between images, this approach leads to an increase in false alarm rate. The binary approach completely excludes background pixels, thus providing more reliable fingerprints.

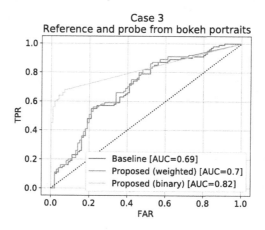

Fig. 8. Case 3: both fingerprint and test images belong to *bokeh portrait* images.

The results clearly show that the performance of PRNU-based source identification is clearly hindered in the presence of the artifacts left by the bokeh mode, if no properly designed countermeasures are taken.

5   Conclusions

In this paper, we observed that, although the sensor pattern noise is still present on HEIC images, it is much more attenuated than in JPEG images, posing serious limitations to its effectiveness in realistic scenarios. Furthermore, we noticed that existing source identification methods are ineffective when images are acquired in *Portrait Mode*. We showed when and how it is possible to address this limitation by removing non-unique artifacts introduced by the device. A possible future extension of this topic is the investigation of whether such artifacts allow to identify the source camera firmware.

References

1. Bianchi, T., Piva, A.: Reverse engineering of double jpeg compression in the presence of image resizing. In: 2012 IEEE International Workshop on Information Forensics and Security (WIFS), pp. 127–132, December 2012. https://doi.org/10.1109/WIFS.2012.6412637
2. Bishop, T.E., Lindskog, A., Molgaard, C., Doepke, F.: Photo-realistic shallow depth-of-field rendering from focal stacks (2017), uS Patent Publication Number US20170070720A1
3. Chen, M., Fridrich, J., Goljan, M., Lukáš, J.: Source digital camcorder identification using sensor photo response non-uniformity. In: Electronic Imaging 2007, pp. 65051G–65051G. International Society for Optics and Photonics (2007)
4. Chen, M., Fridrich, J., Goljan, M., Lukáš, J.: Determining image origin and integrity using sensor noise. IEEE Trans. Inf. Forensics Secur. **3**(1), 74–90 (2008)
5. Goljan, M., Fridrich, J.: Camera identification from cropped and scaled images - art. no. 68190e. Proceedings of SPIE, Media Forensics and Security, March 2008. https://doi.org/10.1117/12.766732
6. Goljan, M., Fridrich, J., Filler, T.: Large scale test of sensor fingerprint camera identification. In: IS&T/SPIE Electronic Imaging, pp. 72540I–72540I. International Society for Optics and Photonics (2009)
7. HEIF related brands and MIME types. https://nokiatech.github.io/heif/technical.html
8. iPhone X market share. https://news.ihsmarkit.com/press-release/technology/iphone-x-led-global-smartphone-unit-shipments-q1-2018-ihs-markit-says
9. ISO/IEC 14496: Information technology. coding of audio-visual objects, part 14: Mp4 file format (2003)
10. ISO/IEC 23008–12:2017: Information technology - high efficiency coding and media delivery in heterogeneous environments - part 12: Image file format (2017)
11. Iuliani, M., Fontani, M., Shullani, D., Piva, A.: Hybrid reference-based video source identification. Sensors **19**(3), 649 (2019). https://doi.org/10.3390/s19030649
12. libheif. https://github.com/strukturag/libheif
13. Lukas, J., Fridrich, J., Goljan, M.: Digital camera identification from sensor pattern noise. IEEE Trans. Inf. Forensics Secur. **1**(2), 205–214 (2006)
14. Mihcak, M.K., Kozintsev, I., Ramchandran, K.: Spatially adaptive statistical modeling of wavelet image coefficients and its application to denoising. In: 1999 IEEE International Conference on Acoustics, Speech, and Signal Processing, 1999. Proceedings, vol. 6, pp. 3253–3256. IEEE (1999)

15. Molgaard, C., Bishop, T.E.: Depth map calculation in a stereo camera system (2017), uS Patent Publication Number US20170069097A1
16. Mondaini, N., Caldelli, R., Piva, A., Barni, M., Cappellini, V.: Detection of malevolent changes in digital video for forensic applications. In: Edward III, J.D., Wong, P.W. (eds.) Security, Steganography, and Watermarking of Multimedia Contents IX, vol. 6505, pp. 300–311. International Society for Optics and Photonics, SPIE (2007). https://doi.org/10.1117/12.704924
17. Otsu, N.: A threshold selection method from gray-level histograms. IEEE Trans. Syst. Man Cybern. **9**(1), 62–66 (1979)
18. Piva, A.: An overview on image forensics. ISRN Signal Process. **2013**, 1–22 (2013)
19. Information property list key reference. https://developer.apple.com/library/archive/documentation/General/Reference/InfoPlistKeyReference/Introduction/Introduction.html

Data-Dependent Scaling of CNN's First Layer for Improved Image Manipulation Detection

Ivan Castillo Camacho and Kai Wang[✉]

Univ. Grenoble Alpes, CNRS, Grenoble INP, GIPSA-lab, Grenoble, France
{ivan.castillo-camacho,kai.wang}@gipsa-lab.grenoble-inp.fr

Abstract. Convolutional Neural Networks (CNNs) have become an effective tool to detect image manipulation operations, *e.g.*, noise addition, median filtering and JPEG compression. In this paper, we propose a simple and practical method for adjusting the CNN's first layer, based on a proper scaling of first-layer filters with a data-dependent approach. The key idea is to keep the stability of the variance of data flow in a CNN. We also present studies on the output variance for convolutional filter, which are the basis of our proposed scaling. The proposed method can cope well with different first-layer initialization algorithms and different CNN architectures. The experiments are performed with two challenging forensic problems, *i.e.*, a multi-class classification problem of a group of manipulation operations and a binary detection problem of JPEG compression with high quality factor, both on relatively small image patches. Experimental results show the utility of our method with a noticeable and consistent performance improvement after scaling.

Keywords: Image forensics · Convolutional Neural Network · First-layer convolutional filter · Image manipulation detection · Stability of variance

1 Introduction

Rapid technology development of cameras to capture digital images comes together with a big suite of software to modify an image. Now changes on an image can be so subtle that noticing them for the naked eye is a difficult task. At the same time, the development of techniques that analyze intrinsic fingerprints in image data, *i.e.*, the image forensics research, is one of the most effective ways to solve challenges related to the authentication of digital images.

Basic image manipulation operations such as median filtering, resampling and noise addition are commonly used during the creation of tampered images. Our objective is to detect traces left by these manipulation operations in an image. It is not a surprise that the current trend is to use Convolutional Neural Network (CNN) to detect image manipulations because of its very good forensic performance. In the meanwhile, image forensics researchers in general agree that

© Springer Nature Switzerland AG 2021
X. Zhao et al. (Eds.): IWDW 2020, LNCS 12617, pp. 208–223, 2021.
https://doi.org/10.1007/978-3-030-69449-4_16

specific design is required in order to build successful CNNs for forensic tasks. In particular, the CNN's first layer needs to be carefully designed so as to extract useful information relevant to the forensic task at hand [1,4,9]. Such relevant information is often believed to be in the so-called image residuals, roughly speaking, in the high-frequency components of the image.

Accordingly, for image manipulation detection, a common choice for the CNN's first layer is high-pass filters. Although satisfying results can be achieved, one important aspect, *i.e.*, the stability of the amplitude of the data flow in CNN, has often been ignored or has not been carefully studied. We have the intuition that after the image data passes through a first layer of high-pass filters, the filters' output becomes a weak signal. This would be detrimental to the training of CNN because the data flow shrinks. In this paper, we show that this signal shrinking indeed exists for first-layer filters generated by several popular initialization algorithms that have been used for detecting image manipulations. In addition, with a proper formulation of the first-layer's convolution operation and based on natural image statistics, we provide an intuitive explanation regarding the signal shrinking and subsequently propose a simple scaling method to enhance the output signal. Experimental results, with different first-layer initializations, CNN architectures and classification problems, show that the proposed scaling method leads to consistent improvement of forensic performance.

The remainder of this paper is organized as follows. In Sect. 2, we briefly review the related work. In Sect. 3 we present an experimental study, as well as an intuitive theoretical explanation, regarding the *variance* of the output signal of CNN's first-layer filters of different initializations. Our proposed data-dependent scaling method is described in Sect. 4. Experimental results are presented in Sect. 5. Finally, conclusions are drawn in Sect. 6.

2 Related Work

Early methods for detecting basic image manipulation operations were based on feature extraction and classifier training. Different handcrafted features were proposed to detect specific and *targeted* manipulation operation, *e.g.*, median filtering, resampling and JPEG compression. Afterwards, researchers focus on the more challenging problem of developing a *universal* method for image manipulation detection. Various methods have been proposed, based on steganalysis features, image statistical model and more recently deep learning.

During the last decade, deep learning methods, including CNNs, have gained outstanding success in a wide range of research problems in the computer vision field. CNNs can learn themselves useful features from given data, effectively replacing the difficult task of handcrafted feature design for human experts. In the recent years, CNNs have also been used to solve image forensic problems. Researchers have noticed a fundamental difference between computer vision tasks and image forensics tasks. The former focuses on the semantic content of images, while the latter often looks for a weak signal representing the difference between authentic and manipulated images. Accordingly, image forensics researchers found that directly applying CNN initialization borrowed from the computer vision field, *e.g.*, the popular Xavier initialization [7], results in rather limited performance in detecting image manipulation operations [1,4]. Different customized CNN initialization algorithms have been proposed to cope better with forensic tasks. The basic idea of these methods are more or less similar, *i.e.*, generating or using a kind of high-pass filters at the CNN's first layer.

SRM (Spatial Rich Model) filters are one popular and effective way to initialize the first layer of CNNs that are used to solve image forensic problems, *e.g.*, the detection of manipulation operations [3], of splicing and copy-move forgeries [10], and of inpainted images [9]. SRM filters, a group of handcrafted high-pass filters originally designed for steganalysis [5], are put at CNN's first layer as initialization and this often leads to very good forensic performances. Indeed, as shown later in this paper, in many cases that we tested, SRM filters outperform other kinds of first-layer filters, especially after the proposed scaling.

Bayar and Stamm [1] proposed a new type of *constrained* filters for the first layer of a CNN designed to detect image manipulation operations. The idea is to constrain the network's first layer to learn a group of high-pass filters. This is realized by normalizing the filters before each forward pass of the CNN training. The normalization consists of two steps: firstly, the center element of filter is reset as -1; secondly, all the non-center elements are scaled so that they sum up to 1. In this way the sum of all filter elements is 0, and the constrained first-layer filter behaves like a high-pass one which is effective in suppressing image content. Recently, Castillo Camacho and Wang [2] proposed an alternative way of initializing CNN's first layer for image manipulation detection. This is essentially an adaptation of the conventional Xavier initialization [7] to the situations where it is required to generate high-pass filters after initialization. This method can generate a set of random high-pass filters to be put at CNN's first layer.

In this paper, we consider four different algorithms for first-layer initialization of CNNs with the application to image manipulation detection: the conventional Xavier initialization from the computer vision community [7], the initialization with SRM filters [5], Bayar and Stamm's constrained filters [1], and Castillo Camacho and Wang's high-pass filter initialization [2]. We show that all the four methods can produce filters which shrink the input signal at their output, and that our proposed data-dependent scaling can noticeably and consistently improve the forensic performance for all the four initialization algorithms when tested on different CNN architectures and forensic problems.

3   Variance of Output of Convolutional Filter

It is demonstrated that the stability of the data flow in CNN, as reflected by the *variance* of the signal in and out a layer, is beneficial for the training of CNN [7,8]. Ideally, the variance of input and output of a layer should be *equal* to each other. In this section, we first show that we can predict the variance of the output of a convolutional filter by using statistics of input signal and elements of the filter. Then, we present observations and understandings regarding the output variance for the four initialization algorithms of convolutional filter which are mentioned in the last section. For the sake of brevity, the four algorithms are hereafter called Xavier [7], SRM [5], Bayar [1], and Castillo [2].

3.1 Formulation

Motivation. We observe potential limitations of the four initialization algorithms and have the intuition that the variance of output of a convolutional filter initialized by them may change substantially compared to the input. SRM [5] and Bayar [1] do not take into account the output variance during the initialization, because the two algorithms put directly third-party SRM filters or normalized high-pass filters at first layer without modelling the relation between input and output. Xavier [7] and Castillo [2] consider the input-output relation and generate pseudo-random filters. These two initialization algorithms are based on a statistical point of view and realized by drawing pseudo-random samples, so in practice properties of initialized filters may differ for different realizations. Therefore, it is interesting and important to experimentally and theoretically study the *actual output variance* for each realization of initialized filter.

Formulation for Computing Output Variance. A convolutional layer used in CNN contains a set of learnable filters (also called *kernels*) [8]. During the forward pass, the kernel moves in a sliding-window manner across the input and computes a weighted sum of the local input data and the kernel. This procedure results in a so-called activation map comprising all the local results computed at every sliding movement of the kernel. Now assume that the kernel contains N scalars denoted by $W = (w_1, w_2, ..., w_N)$, then the local input data involved in the computation also contains N scalars, denoted by $X = (x_1, x_2, ..., x_N)$. It is easy to see that the local output y is simply the dot product of W and X, as:

$$y = \langle W, X \rangle = \sum_{i=1}^{N} w_i.x_i. \tag{1}$$

In Xavier [7] and Castillo [2], both w_i and x_i are assumed as independent random variables. In this paper, we take a *new and more practical* point of view. Since we focus on a proper scaling of a given kernel, we assume that the kernel elements w_i are *known constants*, which can be generated by any initialization algorithm. In addition, we do not consider x_i as independent; instead, we consider them as *mutually correlated* random variables reflecting the natural image statistics [11]. With these assumptions and based on the property of variance of weighted sum of variables, we can compute the variance of the output y as

$$\begin{aligned}
\text{Var}(y) = \text{Var}\left(\sum_{i=1}^{N} w_i.x_i\right) &= \sum_{i=1}^{N}\sum_{j=1}^{N} w_i w_j \text{Cov}(x_i, x_j) \\
&= \sum_{i=1}^{N} w_i^2 \text{Var}(x_i) + 2\sum_{1 \le i < j \le N} w_i w_j \text{Cov}(x_i, x_j).
\end{aligned} \tag{2}$$

Table 1. Considered manipulation operations and their parameters.

Median filtering	$FilterSize = 3$
Gaussian blurring	$StandardDeviation = 0.5$, $FilterSize = 3$
Additive Gaussian noise	$StandardDeviation = 1.1$
Resampling	$ScalingFactor \in \{0.9, 1.1\}$
JPEG compression	$QualityFactor \in \{90, 91, ..., 100\}$

The last expression just divides all the relevant terms into two groups: variance terms and covariance terms of the input signal components $(x_1, x_2, ..., x_N)$.

Furthermore, it is well-known that natural images have approximate *translation invariance* [11], implying that $\text{Var}(x_i), i = 1, 2, ..., N$ are almost identical. In addition, the neighboring pixels are usually highly-correlated [11], which means that $\text{Cov}(x_i, x_j)$ is close to $\text{Var}(x_i)$. We approximate $\text{Var}(x_i)$ by $\text{Var}(x)$, the overall variance of input. Then we have the following approximation of Eq. (2):

$$\text{Var}(y) \approx \text{Var}(x) \left(\sum_{i=1}^{N} w_i^2 + 2 \sum_{1 \leq i} \sum_{<j \leq N} w_i w_j C_{ij} \right), \tag{3}$$

with $C_{ij} = \text{Cov}(x_i, x_j)/\text{Var}(x)$ which are in practice smaller than but very close to 1 for small natural image patches due to high correlation of neighboring pixels. Experimentally the above equation approximates very well the output variance. It also helps us to understand the output variance of popular initialized filters used for manipulation detection, as presented in the remaining of this section.

3.2 Convolutional Filter Initialized with High-Pass Filter

We first consider convolutional filter initialized as each of the 30 SRM filters[1] of shape 5×5 (so here $N = 25$). In order to test on real data for convolutional filter, we take as input 64×64 grayscale image patches generated from the Dresden database [6]. The image manipulation operations that we want to detect are listed in Table 1. We then compute the variance of output of each SRM filter by two different methods: the first one with Eq. (3) and the second one with actual convolution between the input and the filter. Hereafter, we call the first as *covariance-based method* because Eq. (3) is mainly based on the covariance terms $\text{Cov}(x_i, x_j)$ of the input signal components $(x_1, x_2, ..., x_N)$, and we call the second one as *convolution-based method*. For the first method, the covariance terms are estimated from 5×5 small patches (same size as SRM filters) which are randomly extracted from the aforementioned 64×64 Dresden image patches.

The results of $\text{Var}(y)/\text{Var}(x)$, *i.e.*, the ratio of output and input variance, are shown in Fig. 1. We can see that the amplitude of $\text{Var}(y)/\text{Var}(x)$ is very

---

[1] The 30 SRM filters can be found in the `class` of `SrmFiller`, starting from line 347 of this webpage https://github.com/tansq/WISERNet/blob/master/filler.hpp.

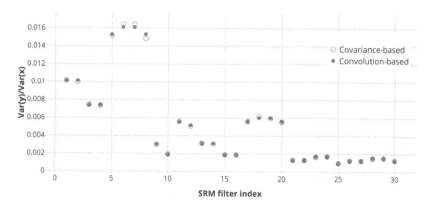

Fig. 1. The value of $\mathrm{Var}(y)/\mathrm{Var}(x)$ for each of the 30 SRM filters obtained by using the covariance-based method of Eq. (3) and the convolution-based method.

small for all 30 SRM filters, which lies basically in a range from 0 to 0.016 with a mean of about 0.005. The output of majority of SRM filters has a variance smaller than 1% of input variance, reflecting the signal shrinkage. It can also be observed that the two methods to obtain output variance give very close results of $\mathrm{Var}(y)/\mathrm{Var}(x)$, implying the coherence of the prediction by Eq. (3) with the practical convolution results.

In order to understand the small output variance for SRM filters, we start from one important property of these high-pass filters. Like many high-pass filters, *e.g.*, Laplacian filter, the sum of all filter elements is equal to 0 for all 30 SRM filters of shape 5×5 (*i.e.*, $N = 25$), which means that we have $\sum_{i=1}^{N} w_i = 0$ (*cf.*, link in footnote (See footnote 1)). It is then easy to deduce that $\sum_{j=1}^{N} w_j \cdot \sum_{i=1}^{N} w_i = \sum_{i=1}^{N} \sum_{j=1}^{N} w_i.w_j = 0$. By dividing the $w_i.w_j$ terms into two groups, we obtain

$$\sum_{i=1}^{N} w_i^2 + 2 \sum_{1 \le i\ <j\le N} \sum w_i w_j = 0. \tag{4}$$

The left-hand side of the above equation is almost same as the term in the bracket of Eq. (3), except that in the above Eq. (4) we replace C_{ij} by 1. As mentioned earlier, for small natural image patches, we have the property that C_{ij} are usually smaller than but very close to 1. This is verified by experiments on Dresden database where the minimum C_{ij} value is 0.9573 for 5×5 small patches. Not surprisingly, this minimum C_{ij} value is attained between two pixel positions which are the farthest from each other within the 5×5 small patch. From the above analysis we can see that the term in the bracket of Eq. (3) is close to 0, which results in a small value of output variance $\mathrm{Var}(y)$ for 30 SRM filters. This intuitively explains the small $\mathrm{Var}(y)/\mathrm{Var}(x)$ values shown in Fig. 1.

Regarding the high-pass filter initialized by Bayar [1] and Castillo [2], we simulated 10,000 filters with both algorithms and calculated the variance of

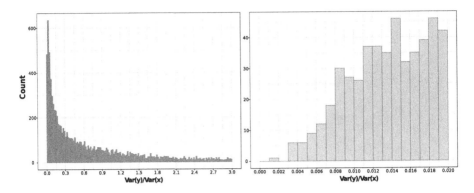

Fig. 2. Two histograms of occurrences of output-input variance ratio $\mathrm{Var}(y)/\mathrm{Var}(x)$ for $10,000$ Xavier filters. Please refer to main text for detailed explanation.

output of simulated filters. We also observe very small values of the ratio of output-input variance, with 0.005 and 0.006 as mean value of $\mathrm{Var}(y)/\mathrm{Var}(x)$, respectively for Bayar and Castillo. Due to space limit, we do not show detailed results, but these small mean values further confirm the behavior that high-pass filters result in small output variance with significant signal shrinkage.

3.3   Convolutional Filter with Xavier Initialization

With curiosity, we also carry out studies for Xavier [7] which generates 5×5 filters filled with pseudo-random samples drawn from a zero-mean uniform distribution. We created $10,000$ Xavier filters using PyTorch and Fig. 2 shows two histograms of occurrences of output-input variance ratio, $i.e.$, $\mathrm{Var}(y)/\mathrm{Var}(x)$: the left one is for the range of 0 to 3 with a bin width of 0.02, while the right one shows detailed occurrences for the first bin of the left histogram for the range of 0 to 0.02 with a bin width of 0.001. We computed the mean value of $\mathrm{Var}(y)/\mathrm{Var}(x)$ for the $10,000$ simulations of Xavier and found that the mean is close to 1 (desired value of Xavier); this is because of a long tail of big values that we do not completely show in Fig. 2. The left histogram of Fig. 2 does not have a peak around 1; instead, the highest occurrences occur near 0. This is a little surprising yet understandable according to Eq. (3). In fact, the elements of Xavier filter are drawn from a zero-mean distribution, so the bracket term in Eq. (3) tends to have a small value. However, due to numerical sampling and in particular considering the relatively low number of 25 pseudo-random samples (for 5×5 filter), it is possible for the bracket term to take big values in certain simulations. Experimentally this bracket term of Eq. (3) can be as big as 13 for some Xavier filters. In addition, from the right histogram of Fig. 2, for Xavier the occurrences of $\mathrm{Var}(y)/\mathrm{Var}(x)$ being very small values, $i.e.$, less than 0.01, is very low: 109 out of $10,000$ simulations ($i.e.$, around 1% probability). In contrast, the majority of this variance ratio is less than 0.01 for high-pass filters as presented in last subsection. We guess that it is still related to the numerical sampling of Xavier because it can be rare to have 25 pseudo-random samples which sum up to a value extremely close to 0.

4   Scaling of Convolutional Filter

From results and analysis in Sect. 3, we can see that the variance of output of convolutional filter initialized by popular algorithms can be significantly smaller than the variance of input. This is particularly true for high-pass filter: the ratio of output-input variance $\text{Var}(y)/\text{Var}(x)$ is usually smaller than 0.01. The output signal after convolution operation substantially shrinks. This can be detrimental to the training of CNN, and as shown later in Sect. 5 the CNN training sometimes fails in such situations.

Using a data-dependent approach (*i.e.*, dependent on input data), we propose a simple yet effective scaling of the first-layer convolutional filter. The idea is to keep the variance stable after scaling for the input and output of any given filter generated by popular initialization algorithms. Corresponding to the two methods to compute output variance in Sect. 3, we propose two different ways to calculate the scaling factor s, as presented below. After obtaining the scaling factor, the elements of the given filter $W = (w_1, w_2, ..., w_N)$ are properly scaled as $\widetilde{W} = s.W$. We then initialize the first-layer filter with the scaled version $\widetilde{W}$.

Covariance-Based Method. From Eq. (3), it can be seen that in order to make $\text{Var}(y)$ and $\text{Var}(x)$ approximately equal to each other, we need to compensate for the effect of the term in the bracket. So the scaling factor is computed as:

$$s = \sqrt{1 \bigg/ \left(\sum_{i=1}^{N} w_i^2 + 2 \sum_{1 \le i < j \le N} w_i w_j C_{ij} \right)}. \qquad (5)$$

In practice, we take random small patches of the same shape of the convolutional filter to be scaled (*e.g.*, 5×5) from a small portion of the training data. We then estimate the variance and covariance terms on these small patches to obtain the values of $C_{ij} = \text{Cov}(x_i, x_j)/\text{Var}(x)$. Afterwards the scaling factor s is computed by using Eq. (5), and at last we obtain the scaled version $\widetilde{W}$ of any given filter W from the considered four initialization algorithms.

Convolution-Based Method. This is a straightforward approach. The output $\widehat{y}$ is computed, for a small portion of the training data $\widehat{x}$ as input, by carrying out the convolution operation. The scaling factor is simply calculated as

$$s = \sqrt{\text{Var}(\widehat{x})/\text{Var}(\widehat{y})}. \qquad (6)$$

From a practical point of view, the covariance-based method might be a slightly better option than the convolution-based method mainly because of its higher flexibility. In fact, for the covariance-based method, the computation of the variance and covariance terms of the input can be performed *only once for any* number of filters for which we want to scale. By contrast, the convolution-based method has to be rerun every time we have a new filter to analyze. Nevertheless, it is worth mentioning that both methods are experimentally quick enough to be used in CNN initialization. The running time is about several seconds, as presented in the next paragraph.

According to our experiments, for a training set of about $100,000$ images of 64×64 pixels from all classes, taking 10% of the training data for the convolution-base method and 10 small patches (*e.g.*, of 5×5 pixels) per image of the 10% training data for the covariance-based method, we achieve a good trade-off between computation time and stability of the result. Using more training data has very small impact on the obtained scaling factor. Even using 100% of the training set results in a change smaller than 0.1%. The amount of time to calculate the scaling factor is less than 3 s per first-layer filter for both methods, on a desktop with Intel® Xeon E5-2640 CPU and Nvidia® 1080 Ti GPU (covariance-based method on CPU and convolution-based method on GPU). This is run for one time before the CNN training. The computation time increases very slowly when having more filters for the covariance-based method, because as mentioned above the variance and covariance terms can be reused. We believe that the computation time of scaling factor is negligible when compared to the typical time required to train a CNN model.

5   Experimental Results

Several experiments are performed in order to test and show the efficiency of our proposed scaling. These experiments consider the four filter initialization algorithms mentioned earlier, two CNN architectures (CNN of Bayar and Stamm [1] and a smaller CNN without fully-connected layer designed by ourselves), and two forensic problems (a multi-class problem of detecting a group of manipulation operations and a binary problem of detecting JPEG compression of high quality factor). For the multi-class problem, we also consider a different number of filters used in the first layer of the CNN of Bayar and Stamm [1]. The implementation and experiments were conducted using PyTorch v1.4.0 with Nvidia® 1080 Ti GPU. The experimental data was created from the Dresden database [6]. Full-resolution Dresden images are split for training, validation and testing with ratio of 3:1:1 and converted to grayscale. Patches of 64×64 pixels were randomly extracted from full-resolution grayscale Dresden images. This relatively small size of image patches makes the forensic problems more challenging.

5.1   Multi-class Problem with CNN of Bayar and Stamm [1]

We first consider the multi-class problem of classifying six different kinds of image patches: the original patches and the five classes of manipulated patches as explained in Table 1. The parameters for the resampling and JPEG compression manipulations are taken randomly from the specified sets in Table 1. The total number of patches in training set is $100,000$ ($\approx$16,667 patches per class), while the number of patches in testing set is 32,000 (≈5,333 patches per class). The number of training and testing samples is same as in [1]. It is worth mentioning that the manipulations and their parameters in this paper are borrowed from [2] and more challenging than those in [1]. The patch size is also smaller than [1]: our patches are of 64 × 64 pixels, while [1] mainly considers 256 × 256 patches.

Table 2. Test accuracy for the multi-class forensic problem (in %, average of 5 runs). The experiments were performed with four initialization algorithms and their scaled versions for first-layer filters of the CNN of Bayar and Stamm [1]. In parentheses is the improvement of scaled version compared to the corresponding original version.

Initialization	Original version	Convolution-based scaling	Covariance-based scaling
Bayar [1] A.	94.19	96.04 (+1.85)	96.02 (+1.83)
Bayar [1] B.		96.15 (+1.96)	96.22 (+2.03)
Castillo [2]	93.71	96.45 (+2.74)	96.42 (+2.71)
SRM [5]	94.39	96.54 (+2.15)	96.55 (+2.16)
Xavier [7]	93.48	94.61 (+1.13)	94.71 (+1.23)

We use the successful CNN architecture of Bayar and Stamm [1] in this set of experiments and initialize the three filters in the CNN's first layer with four different algorithms: Bayar [1], SRM [5], Castillo [2], and Xavier [7]. We carry out 5 runs for each algorithm and the corresponding two scaled versions. For SRM, for each run we randomly select 3 filters from the pool of 30 SRM filters. We compare each original initialization algorithm with their scaled versions obtained with the covariance-based method and the convolution-based method presented in Sect. 4. For fair comparisons, we make sure that for each run the scaled versions share the same "base filters" of the original version before performing scaling. We follow exactly the same training procedure described in [1], including number of epochs, optimization algorithm, learning rate schedule, *etc.*

For Bayar algorithm [1], we have tested two variants of the scaling of the first-layer filters. The first one ("Bayar A.") follows closely the idea of Bayar's original constrained training strategy: we carry out scaling of the normalized high-pass filter at the beginning of each forward pass (please refer to the second last paragraph of Sect. 2 for detail of the normalization procedure proposed in [1]). The second variant ("Bayar B.") is computationally cheaper and less complex: the scaling of normalized high-pass filter is only performed in the initialization, and we no longer impose normalization constraint during training. Our intuition behind the second variant is that with a proper scaling of initialized filters even a free training without the constraint of [1] may provide satisfying performance.

The detection performances in terms of test accuracy (*i.e.*, classification accuracy on testing set) for this multi-class problem are presented in Table 2. The reported results are average of 5 runs with randomness, *e.g.*, different first-layer

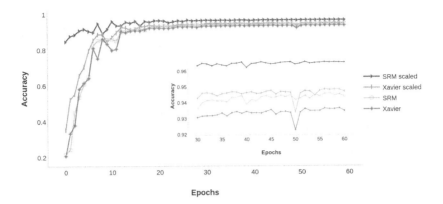

Fig. 3. Curves of test accuracy (average of 5 runs) for the multi-class forensic problem, during the whole 60 training epochs of the CNN of [1]. The curves are for SRM and Xavier, original version and scaled version by the covariance-based method.

"base filters". However, for each run, the "base filters" are the same for the original and scaled versions: original version direct uses these filters, while scaled versions apply proper scaling on the "base filters" and then use the scaled ones.

We can see from Table 2 that the test accuracy of all the initialization algorithms is consistently and noticeably improved after scaling. Improvement of at least 1.13% and as high as 2.74% is obtained. We also observe that the results of the two scaling methods are very close to each other. We checked the computed scaling factors and found that they are indeed almost identical for the two methods. Furthermore, for the two scaling variants of Bayar [1], variant B gives slightly better results, which is also computationally cheaper as it only performs scaling in initialization without enforcing any constraint during CNN training. This implies that with a good initialization after proper scaling, it might not be necessary to impose training constraint. It is worth mentioning that the results of Bayar in Table 2 are in general lower than those reported in [1] because we now consider a more challenging forensic problem with more difficult manipulations and on smaller patches. The results of Castillo in Table 2 are better than those presented in [2]. This may be due to the differences in the number of training epochs (we follow [1] and train with more epochs) and in the adopted optimization algorithm and learning schedule. In addition, the three kinds of high-pass filters (especially SRM) indeed outperform Xavier, before and after scaling. This demonstrates the difference between forensics and computer vision tasks. Nevertheless, the performance of Xavier is also improved after scaling because as analyzed in Sect. 3.3 Xavier can also result in small variance of output. In fact, for Xavier the probability to have $Var(y)$ smaller than half of $Var(x)$ is about 52.20% in our 10,000 simulations.

We also observe that the proposed scaling helps to have quicker increase of forensic performance during CNN training. We show in Fig. 3 curves of test accuracy of SRM and Xavier (average of 5 runs), before and after the covariance-based scaling. It can be observed that the convergence speed is considerably

Table 3. Test accuracy for the multi-class forensic problem (in %, average of 5 runs). We still use the CNN of [1] but change the number of first-layer filters from 3 to 30.

Initialization	Original version	Convolution-based scaling	Covariance-based scaling
Bayar [1] B	94.91	96.11 (+1.20)	96.04 (+1.13)
Castillo [2]	94.11	96.31 (+2.20)	96.32 (+2.21)
SRM [5]	94.37	96.51 (+2.14)	96.49 (+2.12)
Xavier [7]	91.80	96.03 (+4.23)	96.02 (+4.22)

improved for SRM. For both algorithms, the curve of scaled version is always above that of original version during the whole 60 epochs. It is also interesting to notice that the scaled Xavier performs slightly better than the original SRM.

With 30 Filters at First Layer. Next, we present results for the same multi-class problem while changing the number of filters in the first layer of the CNN of [1] to 30. We make this change for two reasons: first, to test our approach with a different number of filters in the first layer; and second, to use all the 30 SRM filters which is a common practice in image forensics, *e.g.*, for detecting splicing and copy-move forgeries [10]. For this scenario we still test the four initialization algorithms but only use variant B for scaled Bayar as it proved to obtain slightly better results while being computationally cheaper. The results are presented in Table 3. Again, we observe that scaling the filters with any of the two methods leads to consistently better test accuracy, with an improvement ranging from 1.13% to 4.23%. We notice from Tables 2 and 3 that after increasing the number of first-layer filters, 1) the original version of high-pass initialization (Bayar, Castillo and SRM) has slightly improved or comparable performance while the accuracy of Xavier decreases; and 2) the scaled version of Bayar, Castillo and SRM has comparable performance with the case of 3 filters while Xavier has noticeable improvement. We guess the reason for the good performance of scaled Xavier may be that with 30 filters there is more chance to have a very good filter which after scaling can improve the result. Understanding these observations is not the focus of our paper, and we plan to conduct further analysis in the future.

5.2 JPEG Binary Problem with CNN of Bayar and Stamm [1]

We notice in the multi-class problem that JPEG compression is the most difficult manipulation to detect. In this section we consider the binary classification between original patches and JPEG compressed patches with parameters in Table 1 (*i.e.*, very high quality factor between 90 and 100). This allows us to test the proposed scaling on a different challenging forensic problem. We use the CNN of Bayar and Stamm [1]. The number of training and testing patches per class is the same as in last subsection. All CNN training settings are kept unchanged.

For this binary problem, we consider two initialization algorithms of Bayar [1] and SRM [5], original and scaled versions (variant B for scaled Bayar). Table 4 presents the obtained results (average of 5 runs). This challenging problem makes the original version of both Bayar and SRM struggle to achieve a good performance. Especially, training of SRM can occasionally fail, leading to accuracy close to random guess. Much better average test accuracy is achieved by scaled versions. For SRM [5], a boost of more than 14% is obtained with scaling.

Table 4. Test accuracy for the binary JPEG forensic problem (in %, average of 5 runs). The experiments were performed with Bayar and SRM, original and scaled versions (variant B for scaled Bayar), on the CNN of [1].

Initialization	Original version	Convolution-based scaling	Covariance-based scaling
Bayar [1] B.	88.27	90.80 (+2.53)	90.80 (+2.53)
SRM [5]	78.24	92.33 (+14.09)	92.44 (+14.20)

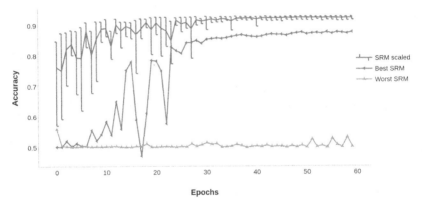

Fig. 4. Curves of test accuracy for the JPEG binary forensic problem: scaled version of SRM (average of 5 runs) with bars of maximum and minimum accuracy at each epoch among 5 runs; and the best run and the worst run of original version of SRM.

We show in Fig. 4 some curves of test accuracy for scaled (covariance-based) and original SRM [5]. The curve of scaled version shows the maximum and minimum test accuracy together with the average at each epoch among the 5 runs. For the original version of SRM we can have very different results. Therefore, we show the best and the worst curves of test accuracy among all the 5 runs. As we can see the worst case does not improve during the whole procedure and the test accuracy remains close to 50%. The difference may come from the randomly selected three first-layer SRM filters in each run (certain SRM filters perform worse than others according to our observation). We would like to mention that for each run, although we select randomly different SRM filters, the

Table 5. Test accuracy for the multi-class and binary problems with our proposed smaller CNN without fully-connected layer (in %, average of 5 runs). The columns of "Bayar" and "SRM" present results of original version. "Scaling-conv" and "Scaling-cov" represent respectively convolution-based and covariance-based scaling method.

Problem	Bayar	Scaling-conv	Scaling-cov	SRM	Scaling-conv	Scaling-cov
Multi-class	95.24	96.17(+0.93)	96.18(+0.94)	95.72	97.06(+1.34)	97.09(+1.37)
Binary	90.56	93.72(+3.16)	93.74(+3.18)	89.85	95.11(+5.26)	95.12(+5.27)

same selected filters are used to carry out comparisons between the original and scaled versions. Therefore, even for filters that result in bad performance for the original version, we can obtain a much better performance after scaling them.

5.3   Multi-class and Binary Problems on a Different Smaller CNN

We then test both the multi-class and JPEG binary problems on a different CNN designed by ourselves. We first describe the architecture of this smaller network. Let Ck(M or A) denote a Convolutional-BatchNorm-Tanh(-MaxPool or -AveragePool) layer with k filters. For the first layer we use Hk which denotes a Convolutional layer with k filters. The architecture of our smaller CNN is H3-C40M-C25M-C20M-C15M-C6A. The first four layers have a kernel size of 5×5 while for the last two layers the kernel size is 1×1. All convolutional stride size is 1. The first layer and the last two layers do not have zero-padding, for the other layers the padding size is 2. This is a network without fully-connected layer. To compare with, the architecture proposed by Bayar and Stamm [1] is H3-C96M-C64M-C64M-C128A-F200-F200-F6, where Fk denotes a fully-connected layer with k neurons and Tanh. The number of learnable parameters of the CNN of [1] is about 337K, while our smaller CNN has about 41K parameters.

Using our smaller CNN, we test both the multi-class and the binary problems on a different CNN architecture. All the data preparation and experimental setting are the same as those described in Sects. 5.1 and 5.2. For this set of experiments, we compare the original and scaled versions of Bayar [1] and SRM [5] (variant B for scaled Bayar). Table 5 presents the obtained results. We can see that in all cases the scaled version leads to improved performance compared to the original version. The improvement of test accuracy goes from 0.93% for multi-class problem with Bayar, to 5.27% for binary problem with SRM.

Our objective in this subsection is to show that with a different CNN, our proposed scaling can still reliably improve the performance for different initialization algorithms and forensic problems. Meanwhile, it can be noticed that performance is better with our smaller CNN when compared to the network of [1]. The understanding of this point is beyond the scope of this paper. Our guess is that the forensic problems and/or the amount of data cope better with the smaller CNN's size (less parameters) and architecture (only comprising convolutional layers without fully-connected layer). The thorough analysis and understanding of the relationships between these factors is one part of our future work.

6 Conclusion

We propose a new and effective scaling approach for adjusting first-layer filters of CNNs used for image manipulation detection. The proposed scaling is computationally efficient and data-dependent (*i.e.*, scaling factor dependent on the input). We also present theoretical and experimental studies which help to understand why the ratio of output-input variance for first-layer convolutional filter can be a (very) small value. Experimental results, with different CNNs, filter initialization algorithms and forensic problems, show that our proposed scaling can consistently improve performance of CNN-based image manipulation detection. Although practically and intuitively we can understand the effectiveness of the proposed filter scaling operation, a rigorous theoretical analysis would be necessary to explain the observed performance improvement. We would like to conduct research on a possible scaling of deeper layers and to extend experiments to more CNNs (*e.g.*, XceptionNet and ResNet), training settings and multimedia security problems. It is also interesting to study the impact of amount of training data and CNN architecture on the forensic performance. One limitation of our work is that we only consider image manipulation detection under ideal laboratory conditions. We plan to carry out relevant studies for more challenging and practical forensic problems, *e.g.*, the detection of malicious and realistic image forgeries and image forensics with mismatch between training and testing data.

Acknowledgments. This work is partially funded by the French National Research Agency (DEFALS ANR-16-DEFA-0003, ANR-15-IDEX-02) and the Mexican National Council of Science and Technology (CONACYT).

References

1. Bayar, B., Stamm, M.C.: Constrained convolutional neural networks: a new approach towards general purpose image manipulation detection. IEEE Trans. Inf. Forensics Secur. **13**(11), 2691–2706 (2018)
2. Castillo Camacho, I., Wang, K.: A simple and effective initialization of CNN for forensics of image processing operations. In: Proceedings of the ACM Workshop on Information Hiding and Multimedia Security, Paris, France, pp. 107–112 (2019)
3. Chen, B., Li, H., Luo, W., Huang, J.: Image processing operations identification via convolutional neural network. Sci. China Inf. Sci. **63**(3) (2020). https://doi.org/10.1007/s11432-018-9492-6
4. Chen, J., Kang, X., Liu, Y., Wang, Z.J.: Median filtering forensics based on convolutional neural networks. IEEE Sig. Process. Lett. **22**(11), 1849–1853 (2015)
5. Fridrich, J., Kodovský, J.: Rich models for steganalysis of digital images. IEEE Trans. Inf. Forensics Secur. **7**(3), 868–882 (2012)
6. Gloe, T., Böhme, R.: The Dresden image database for benchmarking digital image forensics. In: Proceedings of the ACM Symposium on Applied Computing, Sierre, Switzerland, pp. 1584–1590 (2010)
7. Glorot, X., Bengio, Y.: Understanding the difficulty of training deep feedforward neural networks. In: Proceedings of the International Conference on Artificial Intelligence and Statistics, Sardinia, Italy, pp. 249–256 (2010)

8. Li, F.F., Johnson, J., Yeung, S.: Neural networks part 2: setting up the data and the loss (2018). Course notes of Stanford University "CS231n: Convolutional Neural Networks for Visual Recognition". http://cs231n.github.io/neural-networks-2/. Accessed 16 Dec 2020

9. Li, H., Huang, J.: Localization of deep inpainting using high-pass fully convolutional network. In: Proceedings of the IEEE International Conference on Computer Vision, pp. 8301–8310 (2019)

10. Liu, Y., Guan, Q., Zhao, X., Cao, Y.: Image forgery localization based on multiscale convolutional neural networks. In: Proceedings of the ACM Workshop on Information Hiding and Multimedia Security, Innsbruck, Austria, pp. 85–90 (2018)

11. Simoncelli, E.P., Olshausen, B.A.: Natural image statistics and neural representation. Ann. Rev. Neurosci. **24**(1), 1193–1216 (2001)

ISO Setting Estimation Based on Convolutional Neural Network and its Application in Image Forensics

Hui Zeng , Kang Deng, and Anjie Peng[✉]

School of Computer Science and Technology, Southwest University
of Science and Technology, Mianyang, China
penganjie200012@163.com

Abstract. The ISO setting, which is also known as film speed, influences the noise characteristics of output images. As a consequence, it plays an important role in noise based forensics. Whenever the ISO setting information cannot be retrieved from the image metadata, estimating the ISO setting of a probe image from its content is of forensic significance. In this work, we propose a convolutional neural network, called ISONet, for ISO setting estimation. The proposed ISONet can successfully infer the ISO setting both globally (image-level) and locally (patch-level). It not only works on uncompressed images, but also is effective on JPEG compressed images. We apply the ISONet on two typical forensic scenarios, one is the image splicing localization and the other is the Photo Response Non-Uniformity (PRNU) correlation prediction. A series of experiments show that the ISONet can yield a remarkable improvement in both forensic scenarios.

Keywords: ISO setting · Convolutional neural network · Image splicing localization · Photo response non-uniformity · Correlation prediction

1 Introduction

Nowadays, the Internet is overflowing with images captured by digital cameras. At the same time, more and more user friendly software makes image tampering accessible to everybody, threatening the credibility of image contents in insurance claims and courts. To address this challenge, a large number of techniques are proposed to verify the integrity of digital images [1–3]. Different methods are based on different assumptions or different forensic information concerning the questioned image. For example, the works in [4–8] assume that blurring, median filtering, resampling, contrast enhancement, and double JPEG compression is involved in the image tampering, respectively. In [9–11], the authors explore the information of photo response non-uniformity (PRNU), color filter array (CFA) pattern, and JPEG quality factor for forensic purposes. Among various types of forensic information, knowledge of the camera settings during the image acquisition is of our interest.

In practice, the photographers like to set the camera according to the subject and environment. Among the common adjustable settings, the ISO speed can significantly

© Springer Nature Switzerland AG 2021
X. Zhao et al. (Eds.): IWDW 2020, LNCS 12617, pp. 224–236, 2021.
https://doi.org/10.1007/978-3-030-69449-4_17

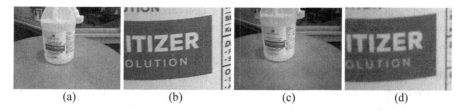

Fig. 1. Images taken under different ISO settings. (a) ISO100, (b) magnified region of (a), (c) ISO3200, (d) magnified region of (c).

affect the characteristics of the output image, e.g., higher ISO speed generally leads to brighter image, at the expense of more noticeable noise [12]. Figure 1 shows two images of similar scene with different ISO settings. It is observed that the ISO3200 image (Fig. 1(c)) is much noisier than the image taken with ISO100 (Fig. 1(a)). This suggests that estimating the ISO setting from the output image should be possible. From the perspective of forensics, estimating the ISO setting is significant for understanding the history of the probe image and furthermore, revealing potential forgeries. First, it is reasonable to assume that the effect of ISO speed on an original image is global. Inconsistences may occur when a forgery is created by splicing two images with different ISO speed together [13]. Second, knowledge of ISO setting is important side information in certain forensic scenarios [14].

The objective of this study is to estimate the ISO setting of a probe image, both globally (image-level) and locally (patch-level), without a reference to the EXIF data. To this end, we propose a convolutional neural network (CNN) [15] to infer the ISO setting from the image alone. The contributions of this work are summarized as following:

1) A CNN model, which is called ISONet, is proposed to estimate the ISO setting of a given image. The proposed ISONet works on both uncompressed images and JPEG compressed images.
2) The proposed ISONet achieves improved image splicing localization performance compared with state-of-the-art noise based methods.
3) We show that PRNU correlation prediction can also benefit from the proposed ISONet.

The rest of this paper is organized as follows. After reviewing the related works in Sect. 2, we provide details of the proposed ISONet in Sect. 3. Experimental study is presented in the fourth section, and the conclusion is drawn in Sect. 5.

2   Prior Work

The last decade witnessed a large number of successful data-driven image forensic methods [5, 11, 16–20]. Rather than presenting a thorough survey of these methods we review a few closely related works. In [5], a median filtering forensic method was proposed by adding a filter layer to the CNN model. In [11], JPEG quality factor was estimated by a CNN with pixel precision. In [16], a camera model fingerprint was extracted with a Siamese network. In [17], a CNN-based model was proposed to improve the sensor pattern noise extraction. In [18], a CNN model was proposed to predict the correlation between an image patch and the reference PRNU.

Inspired by the works mentioned above, we attempt to infer the ISO setting from a probe image for forensic purpose with CNN. To our best knowledge, no previous work in image forensics area focused on ISO setting estimation. However, there are a number of image splicing localization methods based on noise level estimation [13, 21–26]. Considering the fact that a higher ISO setting introduces a higher noise level to the image content, we believe that these noise based methods are most relevant to our work and thus will be compared in Sect. 4.

Another motivation for our work is enhancing the performance of PRNU correlation predictor. PRNU, caused by inhomogeneity of silicon wafers, has been proved to be a reliable fingerprint that provides a link between an image and its source camera. An absence of the PRNU in a certain location of the investigated image can be regarded as a valuable clue of tampering. However, even in a pristine image, the strength of PRNU signal varies significantly over different regions of the image. To reduce false alarms in PRNU based tampering detection, a correlation predictor that can produce an expected PRNU strength is proposed in [27]. In [14], the authors pointed out that the PRNU correlation predictor should be ISO setting specific. Hence, we infer that the correlation predictor can also benefit from the proposed ISONet.

3 ISONet

In this section, we describe the structure, training, and inferring process of the ISONet. Considering that the effect of ISO setting on output images may be significantly altered by JPEG compression, we trained two versions of ISONet. One is for uncompressed images, which is called ISONet-uncompressed, and the other is for JPEG compressed images, called ISONet-JPEG. The two versions of ISONet share exactly the same structure.[1]

3.1 Network Structure

The CNN architecture used in this work is similar to the well-known Very Deep Neural Networks [28], which consist of a convolutional part and a fully connected part. We

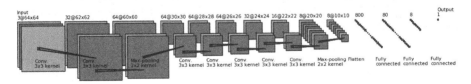

Fig. 2. The structure of the proposed ISONet. The input is a 64 × 64 RGB image patch and the output is a real number indicating the ISO setting.

---

[1] Source code is available at https://github.com/zengh5/ISONet.

choose such a structure based on the following considerations. First, the work of [29] showed that it is possible to infer the local noise level using a stack of convolutional layers with 3 × 3 kernels. Second, as reported in [13], the local noise levels are not determined by ISO speed only, but also affected by local image content. We reasonably believe that the convolutional part can learn the local noise level information and local image content simultaneously. The remaining fully connected part can be used to map the two types of information to ISO speed.

The detail architecture of ISONet is illustrated in Fig. 2. It consists of seven convolutional layers and three fully connected layers. From the first to the last convolutional layer, the number of feature maps is set to 32, 64, 64, 64, 32, 16, and 8, respectively. Two max-pooling layers are inserted after the second convolutional layer and the last layer to reduce the feature dimension. The filter size is 3 × 3 and no padding is used, so the feature maps shrink two pixels after each convolutional layer. The activation function is ReLU for all layers. The input of the network is a 64 × 64 RGB image patch, and the output is a real number related to the ISO setting.

3.2 Training

To train the ISONet-uncompressed, 240 full size uncompressed images with the ISO speed of {100, 200, 400, 800, 1600, and 3200} are randomly selected from the BOSSraw database [30]. The BOSSraw dataset includes 10,000 images from 7 cameras. We try our best to force the selected images to fairly represent all the cameras. Each ISO setting occurs in 40 images. 1064 64 × 64 patches are randomly cropped from each full size image. The rotation and flip based data augmentation is then adopted to generate the final training dataset, which includes 3 × 240 × 1064 = 766,080 patches. For the training target, we define an ISO metric as

$$M_{ISO} = \log_2(ISO\,speed/100) \qquad (1)$$

Such definition reflects the fact that the ISO speed is unevenly distributed in practice, i.e., dense in low ISO region and sparse in high ISO region. The number of epochs is set to 50. The learning rate starts at 1e−3 for the first 10 epochs, then changes

Table 1. Cameras involved in training ISONet-JPEG.

ISO	100	200	400	800	1600	3200
Dataset	DID				Warwick	
Camera	Olympus_C0	Nikon_C0	Olympus_C0	Olympus_C0	Canon 6D	Canon 6D
	Panasonic_C0	Olympus_C0	Olympus_C1	Olympus_C1	Canon 6D MkII	Canon 6D MkII
	Pentax_C0	Olympus_C1	Panasonic_C0	Pentax_C0	Canon 80D	Canon 80D
	Ricoh_C0	Panasonic_C0	Sony_H50_C0	Sony_H50_C0	Fujifilm X_C1	Fujifilm X_C1
		Pentax_C0	Pentax_C0			
		Ricoh_C0	Ricoh_C0			

to 1e−4 for the following 20 epochs, and finally switches to 1e−5 for the remaining 20 epochs. The ADAM algorithm [31] is adopted to optimize ISONet with the MSE loss function.

The training process of the ISONet-JPEG is the same as that of the ISONet-uncompressed, except the used images. The images with ISO speed of {100, 200, 400 and 800} are from the Dresden Image Database (DID) [32]. For the ISO1600 and ISO3200 images, since there is not enough images with these ISO settings in DID, we use the images from the Warwick Image Database [33] instead. All the images are in JPEG format. The involved cameras are listed in Table 1.

3.3 Inference

Inference of the ISO setting of a probe image is simple feed-forward process. For the convenience of subsequent forensic analysis, we estimate the ISO setting both globally (image-level) and locally (patch-level). Probe images are cropped to non-overlapping 64×64 patches $p^{(i)}$, $i = 1, 2, 3, \ldots, N$, and fed into the trained network. The network outputs the estimated ISO metric $\hat{M}_{ISO}^{(i)}$ for each patch. The ISO metric of the whole image is estimated as the median value of the estimated ISO metric of all patches.

$$\hat{M}_{ISO}^{image} = median\left(\hat{M}_{ISO}^{(i)}\right) \qquad (2)$$

4 Experiments

To evaluate the performance of the proposed ISONet, we conducted three experiments as follows. First, the accuracy of the ISO setting estimation on image patches is measured in the baseline experiment. Second, the ISONet-uncompressed is applied in image splicing localization, where images with distinct ISO settings are spliced into a forged image. Third, the ISONet-JPEG is used to infer the ISO setting from JPEG images and enhance the performance of PRNU correlation predictor.

Before moving to the experimental results, it is worthwhile to point out the difference between the forensic senarios investigated in Sects. 4.2 and 4.3. In Sect. 4.2, we restrict ourselves to the image splicing scenario, where a forged image is created by copying a certain region from one source image and pasting into another one (target image). The ISO settings in capturing these two source images are different. In Sect. 4.3, the forged image can be created by any type of tampering operation, e.g., copy-move, splicing, or object removing with the Clone Stamp Tool in Photoshop software. The assumption is that the investigator has the fingerprint and the correlation predictor of the camera at hand. Although the final results can be shown in a similar manner, generally speaking, the forensic strategies for these two scenarios are completely different. As can be seen in the following, our proposed ISONet is helpful in both scenarios.

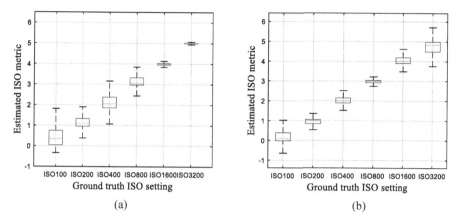

Fig. 3. Performance of ISO setting estimation on 64 × 64 image patches. (a) Uncompressed images, (b) JPEG compressed images.

4.1 Baseline Results

We first test the performance of the ISONet-uncompressed. For each ISO setting, 40 full size images are selected from the BOSSraw dataset and 2,400 64 × 64 patches are cropped from each image, resulting in 96,000 patches for each ISO speed. We assure the test images are not involved in training. The results are shown with box plots in Fig. 3(a), where the red central mark is the median value, and the edges of the box are the 25th and 75th percentiles. It can be observed that the ISONet-uncompressed can infer the ISO setting from the image patches accurately, especially for high ISO cases, e.g. ISO1600 and ISO3200.

We then test the performance of the ISONet-JPEG. The experimental setting is similar to that of the ISONet-uncompressed, except for that the images are from DID and Warwick. In selecting the test images, we assure the cameras are not involved in training. The results are shown in Fig. 3(b), from which it can be observed that the ISONet-JPEG can infer the ISO setting from the JPEG compressed image patches well, except for the ISO3200 images, where the ISO metrics are underestimated and over-lapped with the metrics of ISO1600 images for some cases. Comparing with the results of uncompressed images shown in Fig. 3(a), it seems that the noise characteristics of high ISO images are more blurred by JPEG compression than that of low ISO images.

4.2 Image Splicing Localization

The promising results of ISO setting estimation on image patches suggest that ISONet can be used for image splicing localization when the source images involved in splicing are with distinct ISO settings. Since, to our best knowledge, there is no previous data-

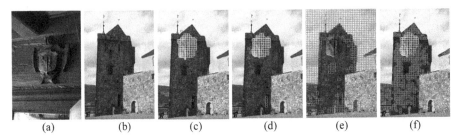

Fig. 4. Image splicing localization. (a) A source image taken with ISO speed = 800, (b) another source image taken with ISO speed = 200, (c) the result with the proposed ISONet, (d) the method in [13], (e) the method in [21], (f) and the method in [23]. The true positives are marked in *green* and the false alarms are marked as *red*. (Color figure online)

driven based method available for comparison we compared the proposed ISONet-uncompressed to state-of-the-art noise based methods [13, 21, 23][2].

Figure 4 shows an image splicing example, in which a badge from one ISO800 image is spliced into one ISO200 image. The original images involved in such a forgery are shown in the Fig. 4(a) and (b). For a fair comparison, all the compared methods are tested on non-overlapping 64×64 blocks, and K-means algorithm is used to cluster these blocks into an original cluster and a spliced cluster based on the forensic metrics obtained with corresponding methods. The localization results of the proposed method, the method [13], the method [21] and the method [23] are shown in Fig. 4(c), (d), (e), and (f), respectively. While the proposed ISONet and the method of [13] are able to provide meaningful clues, the proposed method has fewer false alarms.

We also make a quantitative comparison between the proposed method and the compared methods on uncompressed images from the BOSSraw database. 40 ISO100 images are fixed as the background. A randomly selected 1024×1024 region from an image with ISO speed $\in \{200, 400, 800, 1600, 3200\}$ is spliced into each background image to generate a forged image. We use the block detection accuracy (BDA) and the block false positive (BFP) to evaluate the splicing localization performance, as in [13, 21] and [23]. BDA is the probability of image blocks in the spliced region that are correctly detected and BFP is the probability of image blocks in the original region that are falsely detected. Figure 5 shows the average BDA/BFP rates over 40 images when the spliced regions are taken with different ISO settings. The red square lines represent the ISONet, the green circle marked lines represent the method in [21], the blue rhombus marked lines represent the method in [23], and the magenta star marked lines represent the method in [13]. The ISONet consistently performs the best among the compared four methods, especially when ISO speed difference between the spliced region and the background is small. For example, when the spliced regions are from ISO400 images, the ISONet achieves BDA = 99.0% and BFP = 7.4%, while the method in [21] achieves BDA = 67.9% and BFP = 16.3%, the method in [23] achieves

---

[2] The implementation of [13] is available in https://github.com/zengh5/Exposing-splicing-sensor-noise, and the implementations of [21, 23] are available in https://github.com/MKLab-ITI/image-forensics/tree/master/matlab_toolbox.

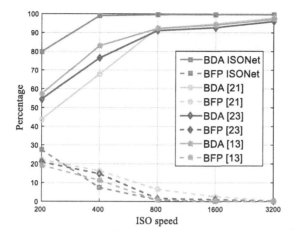

Fig. 5. BDA/BFP rates comparison for image splicing localization. The forged images are generated by splicing 1024 × 1024 blocks of different ISO settings into ISO100 images.

BDA = 76.5% and BFP = 14.7%, and the method in [13] achieves BDA = 83.0% and BFP = 11.3%.

It is worthwhile to point out that all the compared noise based methods, as well as the ISONet, are not robust to JPEG compression after image splicing. However, the proposed ISONet-JPEG can be used to reveal the forgery (no double compression) when the source images are in JPEG format.

In addition to testing on our own generated spliced images, we also tested the proposed model on the Realistic Tampering dataset [34]. For most of the tampering cases in this dataset, the involved source images are captured with very low ISO setting, e.g. ISO100, which makes it impossible to localize the forgery by local ISO estimation. Even so, the proposed model can give valuable clues for some forgery cases. Figure 6 shows such an example. For the forgery case shown in Fig. 6(a), a football is spliced into the lawn background. The background image seems taken with ISO100 and the patch containing the football may be from a high ISO image, e.g., ISO400 or ISO800.

4.3 PRNU Correlation Prediction

In PRNU based forgery localization, a PRNU correlation predictor is used to predict the PRNU strength of a given image patch, which is essential for reducing the false

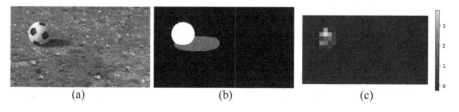

(a)   (b)   (c)

Fig. 6. Local ISO number estimation result for a tampering case from [34]. (a) Tampered image, (b) ground truth mask, (c) block-wise ISO number estimation result with ISONet.

Table 2. Coefficient of determination r^2 of the correlation predictions obtained by different correlation predictors. By choosing the predictor according to the estimated ISO using ISONet, the value of r^2 can be significantly improved over that of the *mixed* predictor.

	Matching			Mixed			Estimated		
ISO	100	200/400	800	100	200/400	800	100	200/400	800
r^2	0.700	0.314	0.111	0.584	0.240	0	0.682	0.314	0.111

alarms [27]. The authors of [14] pointed out that the correlation predictor should be ISO specific. In this subsection, we show how the proposed ISONet can enhance the performance of correlation predictor, and furthermore, enhance the performance of the PRNU based forgery localization.

First, we trained four versions of correlation predictors for the Pentax_OptioA40 camera model in DID: *ISO100*, *ISO200/400*, *ISO800*, and *mixed ISO*. The *ISO100* and *ISO800* correlation predictors are trained with 20 images of the corresponding ISO settings. The number of ISO200 or ISO400 images is insufficient in this camera model for training correlation predictors separately, thus we combine 10 ISO200 images and 10 ISO400 images to train one correlation predictor denoted as *ISO200/400*. The *mixed* correlation predictor is trained with 20 images randomly chosen from the 60 images used for *ISO100*, *ISO200/400*, and *ISO800*. The standard 15D content based features as well as correlations are computed on image patches of 128×128 pixels. The least square estimator is used to obtain the predictors as in [27]. Then, 10,000 128×128 patches of each ISO setting are used to test the performance of the correlation predictors. As in [34], we use the coefficient of determination (r^2) to evaluate the quality of the obtained correlation predictors. This coefficient, ranging from 0 to 1, is the proportion of the variance in the dependent variable that is predictable from the independent variables. The higher the r^2 value, the higher quality the correlation predictor is. Table 2 compares r^2 when the matched correlation predictor (the ISO setting of the test image is the same with the correlation predictor), the *mixed* correlation predictor, and the predictor chosen according to the ISONet-JPEG are used for prediction. It is observed that the predictors chosen according to the ISONet-JPEG show superior performance over the *mixed* correlation predictor, and perform similar to the matched correlation predictors. With the help of the ISONet-JPEG, for most test images the matched correlation predictor works without an error, except for a few ISO100 images that are wrongly classified as ISO200 images.

To provide a more intuitive example, we examine the influence of the correlation predictor in PRNU based forgery localization on a tampered image. The original image shown in Fig. 7(a) is taken with the setting of ISO400. The tampered image (Fig. 7(b)) is manipulated by copying a ship to the same image in Adobe Photoshop. The Standard Localization Algorithm from [34] (Algorithm 1 of [34]) is applied to the tampered image.[3] We set the window size $\omega = 129$, analysis stride $\Delta\omega = 64$, and threshold $\tau = 0.7$ for the Standard Localization Algorithm. Figure 7(c) shows the result without correlation predictor. Alternatively, a fixed threshold based on peak-to-correlation-energy (PCE) [35] is used, i.e., the patches whose PCE values are lower than a given

---

[3] Source code is available at https://github.com/pkorus/multiscale-prnu.

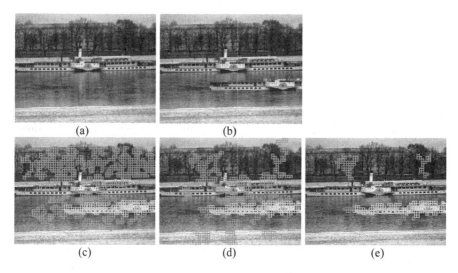

Fig. 7. PRNU based forgery detection results of an ISO400 image. (a) Original image, (b) forged image, (c) without correlation predictor, a fixed threshold PCE = 2 is used, (d) with a mismatched *ISO100* correlation predictor, (e) with the *ISO200/400* correlation predictor chosen according to the proposed ISONet.

threshold t are declared to be suspicious. Here we experimentally set $t = 2.$[4] A lot of false alarms occur in the area of trees. This is due to the fact that it is hard to extract the PRNU signal from the textual area. Figure 7(d) shows the result when the *ISO100* correlation predictor (mismatched) is used. Compared to Fig. 7(c), the false alarms in the area of trees have been suppressed because the weak PRNU strength in these areas has been taken into consideration by the correlation predictor. However, there are false alarms in the pristine ship and the beach at the bottom of the image caused by the overestimation of the *ISO100* correlation predictor. In the proposed scheme, the ISO metric of the tampered image is first estimated with the ISONet-JPEG, which outputs 2.093. Then the *ISO200/400* correlation predictor is chosen according to the definition of the ISO metric in (1) for forgery detection. Figure 7(e) shows the localization result with fewer false alarms compared to Fig. 7(c) and (d).

To summarize, both the correlation predictor and the PRNU based forgery localization can benefit from the proposed ISONet.

5   Conclusion

In this paper, a CNN-based model is proposed to estimate the ISO setting of a given image both globally and locally. The estimated ISO setting is a kind of important information in the image forensics. When the ISONet is used for image splicing

---

[4] For an ISO400 image of JPEG format, the PRNU signal is very weak even in the pristine area. Thus, the threshold here is much lower than that used in camera identification.

localization, it achieves superior performance to state of the art noise based methods. We also show that the PRNU correlation prediction can benefit from the proposed ISONet. The application of the ISONet may span beyond the scenarios discussed in this paper. For example, the estimated ISO speed may be useful side information for steganalysis.

Given the prospective performance reported in this work, it is worthwhile to pointed out the generalizability of the ISONet may not satisfy enough for unknown camera models, especially in the case of ISONet-JPEG. We hypothesize that it is due to different camera models compress images with different quality, which makes JPEG images taken with the same ISO settings from different camera models have different noise characteristics. Inferring the ISO settings that never seen in training, such as ISO160, is also of our interest in future.

Acknowledgment. We would like to thank Dr. M. Goljan for helping us in revising this manuscript, to thank the authors of [34] for their open source codes, and to thank the authors of [33] for sharing their high ISO image dataset. We would also like to thank the anonymous reviewers for their helpful suggestions. This work was supported by NSFC (grant no. 61702429), China Scholarship Council (no. 201908515095), and the Research Fund for the Doctoral Program of Southwest University of Science and Technology University (grant no. 18zx7163).

References

1. Stamm, M.C., Wu, M., Liu, K.J.R.: Information forensics: an overview of the first decade. IEEE Access **1**, 167–200 (2013)
2. Böhme, R., Kirchner, M.: Media forensics. In: Katzenbeisser, S., Petitcolas, F. (eds.) Information Hiding, pp. 231–259. Artech House (2016)
3. Korus, P.: Digital image integrity – a survey of protection and verification techniques. Digit. Signal Process. **71**, 1–26 (2017)
4. Bahrami, K., Kot, A.C., Li, L., Li, H.: Blurred image splicing localization by exposing blur type inconsistency. IEEE Trans. Inf. Forensics Secur. **10**(5), 999–1009 (2015)
5. Chen, J., Kang, X., Liu, Y., Wang, Z.J.: Median filtering forensics based on convolutional neural networks. IEEE Signal Process. Lett. **22**(11), 1849–1853 (2015)
6. Popescu, C., Farid, H.: Exposing digital forgeries by detecting traces of resampling. IEEE Trans. Signal Process. **53**(2), 758–767 (2005)
7. Lin, X., Li, C.T., Hu, Y.: Exposing image forgery through the detection of contrast enhancement. In: Proceedings of IEEE International Conference on Image Processing, pp. 4467–4471 (2013)
8. Bianchi, T., Piva, A.: Image forgery localization via block-grained analysis of JPEG artifacts. IEEE Trans. Inf. Forensics Secur. **7**(3), 1003–1017 (2012)
9. Lukas, J., Fridrich, J., Goljan, M.: Digital camera identification from sensor pattern noise. IEEE Trans. Inf. Forensics Secur. **1**, 205–214 (2006)
10. Jeon, J.J., Shin, H.J., Eom, I.K.: Estimation of Bayer CFA pattern configuration based on singular value decomposition. EURASIP J. Image Video Process. **2017**(1), 1–11 (2017). https://doi.org/10.1186/s13640-017-0196-z
11. Uchida, K., Tanaka, M., Okutomi, M.: Pixelwise JPEG compression detection and quality factor estimation based on convolutional neural network. In: Proceedings of IS&T International Symposium on Electronic Imaging, p. 276 (2019)

12. Petteri, O.M.: Dependence of the parameters of digital image noise model on ISO number, temperature and shutter time. Project work report (2008). https://www.cs.tut.fi/~foi/MobileImagingReport_PetteriOjala_Dec2008.pdf

13. Zeng, H., Peng, A., Lin, X.: Exposing image splicing with inconsistent sensor noise levels. Multimed. Tools Appl. **79**(35–36), 26139–26154 (2020). https://doi.org/10.1007/s11042-020-09280-z

14. Quan, Y., Li, C.: On addressing the impact of ISO speed upon PRNU and forgery detection. IEEE Trans. Inf. Forensics Secur. **16**, 190–202 (2021)

15. LeCun, Y., Bottou, L., Bengio, Y., Haffner, P.: Gradient based learning applied to document recognition. Proc. IEEE **86**(11), 2278–2324 (1998)

16. Cozzolino, D., Verdoliva, L.: Noiseprint: a CNN-based camera model fingerprint. IEEE Trans. Inf. Forensics Secur. **15**, 144–159 (2020)

17. Kirchner, M., Johnson, C.: SPN-CNN: boosting sensor-based source camera attribution with deep learning. In: 2019 IEEE International Workshop on Information Forensics and Security, pp. 1–6 (2019)

18. Chakraborty, S.: A CNN-based correlation predictor for PRNU-based image manipulation localization. In: Proceedings of IS&T International Symposium on Electronic Imaging, p. 078 (2020)

19. Rao, Y., Ni, J.: A deep learning approach to detection of splicing and copy-move forgeries in images. In: 2016 IEEE International Workshop on Information Forensics and Security, pp. 1–6 (2016)

20. Bayar, B., Stamm, M.: A deep learning approach to universal image manipulation detection using a new convolutional layer. In: ACM Workshop on Information Hiding and Multimedia Security, pp. 5–10 (2016)

21. Mahdian, B., Saic, S.: Using noise inconsistencies for blind image forensics. Image Vis. Comput. **27**(10), 1497–1503 (2009)

22. Lyu, S., Pan, X., Zhang, X.: Exposing region splicing forgeries with blind local noise estimation. Int. J. Comput. Vision **110**(2), 202–221 (2014)

23. Zeng, H., Zhan, Y., Kang, X., Lin, X.: Image splicing localization using PCA-based noise level estimation. Multimed. Tools Appl. **76**(4), 4783–4799 (2016). https://doi.org/10.1007/s11042-016-3712-8

24. Yao, H., Cao, F., Tang, Z., Wang, J., Qiao, T.: Expose noise level inconsistency incorporating the inhomogeneity scoring strategy. Multimed. Tools Appl. **77**(14), 18139–18161 (2017). https://doi.org/10.1007/s11042-017-5206-8

25. Zhu, N., Li, Z.: Blind image splicing detection via noise level function. Signal Process: Image Commun. **68**, 181–192 (2018)

26. Zhang, D., Wang, X., Zhang, M., Hu, J.: Image splicing localization using noise distribution characteristic. Multimed. Tools Appl. **78**(16), 22223–22247 (2019). https://doi.org/10.1007/s11042-019-7408-8

27. Chen, M., Fridrich, J., Goljan, M., Lukas, J.: Determining image origin and integrity using sensor noise. IEEE Trans. Inf. Forensics Secur. **3**(1), 74–90 (2008)

28. Simonyan, K., Zisserman, A.: Very deep convolutional networks for large-scale image recognition. In: Proceedings of the International Conference on Learning Representations (2015). https://arxiv.org/abs/1409.1556

29. Guo, S., Yan, Z., Zhang, K., Zuo, W., Zhang, L.: Toward convolutional blind denoising of real photographs. In: Proceedings of the IEEE Conference on Computer Vision and Pattern Recognition, pp. 1712–1722 (2019)

30. Bas, P., Filler, T., Pevný, T.: "Break our steganographic system": the ins and outs of organizing BOSS. In: Filler, T., Pevný, T., Craver, S., Ker, A. (eds.) Information Hiding. IH 2011. Lecture Notes in Computer Science, vol. 6958. Springer, Heidelberg (2011). https://doi.org/10.1007/978-3-642-24178-9_5.

31. Kingma, D.P., Ba, J.L.: ADAM: a method for stochastic optimization. In: Proceedings of the International Conference on Learning Representations, p. 15 (2015)

32. Gloe, T., Bohme, R.: The Dresden image database for benchmarking digital image forensics. J. Digit. Forensic Pract. 3(2–4), 150–159 (2010)

33. Quan, Y., Li, C.-T., Zhou, Y., Li, L.: Warwick image forensics dataset for device fingerprinting in multimedia forensics. In: IEEE International Conference on Multimedia and Expo (ICME), pp. 1–6 (2020)

34. Korus, P., Huang, J.: Multi-scale analysis strategies in PRNU-based tampering localization. IEEE Trans. Inf. Forensics Secur. 12(4), 809–824 (2017)

35. Goljan, M.: Digital camera identification from images–estimating false acceptance probability. In: Kim, H.J., Katzenbeisser, S., Ho, A.T.S. (eds.) Digital Watermarking. IWDW 2008. Lecture Notes in Computer Science, vol. 5450. Springer, Heidelberg (2009). https://doi.org/10.1007/978-3-642-04438-0_38

A Hybrid Loss Network for Localization of Image Manipulation

Qilin Yin[1], Jinwei Wang[1,2,3(✉)], and Xiangyang Luo[2]

[1] Nanjing University of Information Science and Technology and Engineering Research Center of Digital Forensics, Ministry of Education, Nanjing 210044, China
wjwei_2004@163.com
[2] State Key Laboratory of Mathematical Engineering and Advanced Computing, Henan 450001, China
[3] Shanxi Key Laboratory of Network and System Security, Xidian University, Xi'an 710071, China

Abstract. With the development of information security, localization of image manipulations havs become a hot topic. In this paper, a hybrid loss network is proposed for the manipulated image forensics. First, the patch prediction module extracts corresponding features representing the discrepancy of the tampered region boundaries. Then, these features are constrained by the Pixel Normalization, thereby improving the classification performance for the tampered patches. Finally, multi-scale patch prediction masks and semantic information are fused to segment out the tampered regions. The experiments demonstrate the proposed model can achieve high performance.

Keywords: Image manipulation localization · Pixel normalization · Hybrid loss network

1 Introduction

With the rapid development of digital media technology, the emergence of a large number of image editing software enables people to easily manipulate the content of the image, which leads to a sharp decline in the authenticity of the image, resulting in a serious reduction in the credibility of the image. The frequently tampered images have caused serious adverse consequences in many aspects of our life, such as military, judicial, media, etc. [26]. To address this threat, image

This work was jointly supported by the National Natural Science Foundation of China (Grant No. 61772281, 61702235, 62072250, U1636117, and U1636219), in part by the National Key R&D Program of China (Grant No. 2016YFB0801303 and 2016QY01W0105), in part by the plan for Scientific Talent of Henan Province (Grant No. 2018JR0018), in part by Postgraduate Research & Practice Innvoation Program of Jiangsu Province (Grant No. KYCX20_0974) and the Priority Academic Program Development of Jiangsu Higher Education Institutions (PAPD) fund.

X. Zhao et al. (Eds.): IWDW 2020, LNCS 12617, pp. 237–247, 2021.
https://doi.org/10.1007/978-3-030-69449-4_18

manipulation detection and localization have received extensive attention, which aims to recognize the tampered regions.

According to whether the content of the image has been changed or not, the image forgery techniques can be broadly divided into two categories [16]: (1) content-preserving, and (2) content-changing. Content-preserving manipulations (e.g., blur, compression, and enhancement) only change the visual quality of the image, but do not affect the underlying semantic information of the image. However, content-changing manipulations (e.g., splicing, copy-move, and removal) can alter the content of the image arbitrarily, which is also our focus.

In contrast to image object detection which aims to recognize all foreground objects, the image manipulation detection which aims to localize specific regions (including foreground object, or background, or even an erased area), is more difficult and challenging. Depending on the characteristics, most prior works use handcrafted or predetermined features such as JPEG compression artifacts [1, 11], Camera Filter Array (CFA) pattern [4,8], edge inconsistencies [21], and local noise pattern [6,19,25] to segment out the tampered region in an image. Nevertheless, most of these methods focus on one specific type of manipulation resulting in their poor generalization ability. With the great success of deep learning in computer vision, a significant number of deep learning based methods have been proposed to address earlier issues. Although these methods improve the generalization ability of models, they usually rely on heavy, time-consuming pre- and/or post-processing and are still limited to predetermined features [3, 16,24,27].

Focusing on the above investigation, we proposed a unified neural network architecture for image manipulation localization. The proposed network is an end-to-end, multi-task model without pre- and/or post-processing, which integrates the classification of the large, middle, and small patch. The whole coarse-to-fine manipulation localization process can be divided into two stages: 1) Stage-1, the classification process of multi-scale patches cooperates to localize the tampered region from coarse to fine. 2) Stage-2, multi-scale semantic content is introduced into image segmentation to eliminate patch-effect and obtain better pixel-wise localization results.

The main contributions of the present study are as follows:

- A unified hybrid loss network without any pre- and/or post-processing is proposed, which can simultaneously realize the detection of multi-scale tampered patches and the pixel-wise segmentation of tampered regions.
- A patch prediction module based on statistical property differences is proposed, thereby remaining the spatial position relationship between patches intactly and classifying the tampered patches effectively.
- A new regularization method called Pixel Normalization (Pixel Norm) is proposed. This method can alleviates the influence of image content on feature extraction and improve the detection performance of the proposed model.

The rest of this paper is organized as follows: Sect. 2 discusses the related work, Sect. 3 elaborates the proposed model, Sect. 4 details experiments, and Sect. 5 presents conclusions.

2   Related Works

2.1   Traditional Image Processing Based Methods

Before the emergence of deep learning based approaches, the traditional image forensics method was always the first choice of image forgery detection. These methods always take advantage of some physical characteristics as the foundation for detection/localization, including frequency domain characteristics [12,22], artifacts left by JPEG compression [5,10,22], noise pattern [14,15,19], and CFA pattern [20]. Specifically, under the assumption that the tampered regions and the authentic regions have undergone different JPEG compressions, the tampered regions by the analysis of the compression errors through different JPEG compression qualities was detected [10]. In [8], a Gaussian Mixture model was proposed to classify CFA present areas (authentic areas) and CFA absent areas (tampered areas). However, CFA based methods only apply to tampered regions and non-tampered regions from different cameras. Mahdian *et al.* [15] modeled the local noise to localize the boundary of the tampered regions.

2.2   Deep Learning Based Methods

In recent years, inspired by the success of deep learning techniques in the computer vision field, a number of deep learning based methods have been proposed to address the forgery detection. In contrast to traditional image processing based methods, deep learning approaches can extract multi-hierarchical features, which is conducive to improving the generalization ability of models. MFCN [21] is a multi-task fully convolutional network that is simultaneously trained on the edge binary mask and the tampered region mask. Tampered region edge detection is helpful to reduce the error detection rate of non-tampered pixels and can yield finer localization of the spliced region. Bappy *et al.* [3] proposed an LSTM based network to find out the small patches on the boundaries of the tampered regions. Then, this network was jointly trained with pixel-wise segmentation for finer localization. However, the scale of the patch is difficult to determine. If it is too large, the localization result is very coarse. If it is too small, the information provided by the patch is less, resulting in the small patch is difficult to detect. Authors in [27] utilized the mainstream object detection network, supplemented by some specific features to identify fake objects, but this method fails to achieve pixel-wise segmentation. The fusion of spatial features and frequency domain feature is adopted in [16] for detecting manipulated image regions. Work in [24] formulated the image manipulation detection task as a local anomaly detection problem. A learnable decision function is proposed to establish the mapping between the difference of the local feature and the corresponding forgery label.

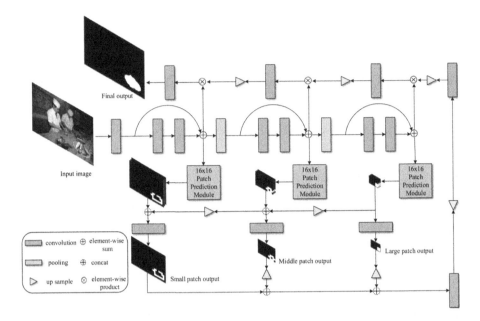

Fig. 1. Overview of the proposed method.

3   Proposed Method

3.1   Overview

The logical structure of the proposed method is shown in Fig. 1. Firstly, the input will be filtered by three residual blocks. Pooling operations following the convolution operations are introduced into the model to compress the data dimensions. Then, the output feature maps of each residual block are sent to the patch prediction module we proposed to perform the multi-scale localization of the boundaries of the tampered region. Although the size of the extracted patch in the prediction module is 16 × 16, these feature maps have undergone multiple pooling operations. The size of 16 × 16 patch extracted by the patch prediction module mapping to the original image is 16, 32, and 64, respectively. Finally, the patch prediction masks will undergo a series of convolution and up-sample operations to yield the final pixel-wise manipulation localization mask.

3.2   Patch Prediction Module

For a forgery image, even with careful inspection, it is difficult to recognize the tampered regions, because advanced manipulation techniques cannot leave any visible traces. However, the manipulation must distort the statistical properties of an image, especially in the boundaries that tampered and non-tampered regions share. Although these differences are indistinguishable to humans, we can use deep learning techniques to represent the statistical properties of the

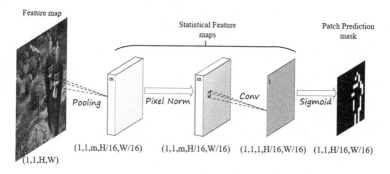

Fig. 2. Diagrammatic overview of the patch prediction process of a single feature map.

images and further capture these differences. Thus, we propose a patch prediction module to learn the correlation between the patches extracted from each feature map.

As shown in Fig. 2, the patch prediction process of a feature map is considered as an example. Firstly, the dimension of a feature map is $(1, 1, H, W)$, 16×16 pooling operation is used to simultaneously extract the patch of the image and calculate the m statistical features of each patch. In this work, the most common statistics, maximum, minimum, mean and variance, are selected as indicators to represent the statistical properties of the extracted patches. After the pooling operation, m statistical feature maps have been obtained, the dimension of which is $(1, 1, m, H/16, W/16)$. It is worth noting that, in addition to batch (N), channel (C), height (H), and weight (W), the dimension of the feature map has been increased by one dimension, the number of statistical feature (S). Then, the statistical feature maps are normalized by the proposed Pixel Normalization. Finally, the patch prediction mask can be achieved by a convolutional layer with sigmoid activation function.

3.3 Pixel Normalization

In the prior works, a family of feature normalization methods have been proposed for different scenarios and tasks, including Batch Norm [9], Layer Norm [2]. They all obey a general formulation of feature normalization:

$$\hat{x}_i = \frac{1}{\sigma_i}(x_i - \mu_i) \tag{1}$$

where x is the feature computed by a layer, and i is an index. In the case of 2D images, $i = (i_N, i_C, i_H, i_W)$ is a 4D vector indexing the features in (N, C, H, W) order, where N is the batch axis, C is the channel axis, and H and W are the spatial height and weight. μ_i and σ_i in Eq. 1 are the mean and standard deviation. They can be calculated by the following formulas:

$$\mu_i = \frac{1}{m} \sum_{k \in S_i} x_k \quad \sigma_i = \sqrt{\frac{1}{m} \sum_{k \in S_i} (x_k - \mu_i)^2 + \varepsilon} \tag{2}$$

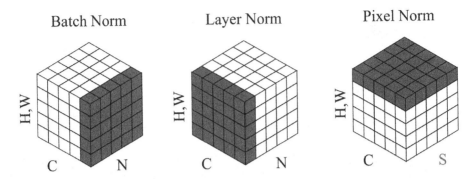

Fig. 3. Normalization methods. Each subplot shows a feature map tensor, with N as the batch axis, C as the channel axis, S as the statistics axis, and (H, W) as the spatial axes. The pixels in blue are normalized by the same mean and variance, computed by aggregating the values of these pixels. (Color figure online)

where ε is an arbitrarily small value. S_i is a set of pixels which are used to compute the mean and standard deviation. m is the size of this set. The difference between these methods mainly lies in how the set S_i is defined, discussed as follows Fig. 3:

In **Batch Norm**, the set S_i is defined as:

$$S_i = \{k | k_C = i_C\} \tag{3}$$

where k_C (and i_C) denotes the sub-index of k (and i) along the axis C. That is to say that the pixels sharing the same channel index are normalized together. Batch Norm computes the μ and σ along the (N, H, W) axes. In **Layer Norm**, the set S_i is defined as:

$$S_i = \{k | k_N = i_N\} \tag{4}$$

meaning that Layer Norm computes the μ and σ along the (C, H, W) axes.

Pixel Normalization (Pixel Norm). As analyzed in the previous section, in the terms of the features received by Pixel Norm, $i = (i_N, i_C, i_S, i_H, i_W)$ is a 5D vector indexing the features in (N, C, S, H, W) order, where S is the statistics axis. So, the Pixel Norm actually performs feature normalization on n features with (C, S, H, W) order, respectively. Formally, in Pixel Norm, the set S_i is defined as:

$$S_i = \{k | k_{H,W} = i_{H,W}\} \tag{5}$$

meaning that Pixel Norm computes the μ and σ along the (C, S) axes. The computation of Pixel Norm is illustrated in Fig. 3 (rightmost).

For the features with (C, S, H, W) order, each pixel along the (C, S) axes represents the corresponding statistical feature extracted from a patch in an image. Due to each patch containing different contents, there may be great differences between the statistics for each patch, *i.e.*, the magnitude of statistics, which may affect the performance of the patch classification. To avoid this problem, the Pixel Norm is proposed to normalize each pixel along the (C, S) axes, making the statistical properties of each patch tend to be similarly distributed. Because of the share of the convolution kernel parameters, similar statistical properties distribution is more conducive to the patch detection.

3.4   Training Loss

The proposed model has a total of four outputs. We used different cross-entropy losses for different branch learning. The total loss function of the proposed model is defined as:

$$L_{total} = \alpha_1 F_{large} + \alpha_2 F_{mid} + \alpha_3 F_{small} + \alpha_4 C_{seg} \qquad (6)$$

where F_{large} is the Focal loss [13] of the large patch classification, F_{mid} is the Focal loss of the middle patch classification, F_{small} is the Focal loss of the small patch classification, and L_{seg} is the standard cross-entropy for the final pixel-wise segmentation loss. α_1, α_2, α_3, and α_4 are the pre-defined hyper-parameter. In this work, they are defined as 0.6, 0.1, 0.1, and 0.2, respectively. Finally, the four losses are summed together to produce the loss function for the whole model.

4   Experiments

In this section, extensive experiments are carried out to demonstrate the effectiveness of each part of the proposed model. We evaluate the proposed model on four image forensics benchmarks and compare the results with state-of-the-art methods.

4.1   Implementation Settings

In data preparation, the input image size is resized to $(256, 384, 3)$. Image flipping is used for the data augmentation. According to the existing ground truth masks, we can easily generate the patch ground truth mask. The batch size of the proposed model is 16. The whole network is optimized by Adam with a learning rate set at 1e−3 (pre-train) and 5e−4 (fine-tune). The training ends at epoch 200. All experiments are performed on a single NVIDIA 2080Ti GPU.

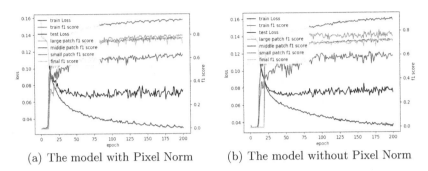

(a) The model with Pixel Norm   (b) The model without Pixel Norm

Fig. 4. The F1 score and loss of the model with and without Pixel Norm

4.2 Experimental Analysis

We use pixel-level F1 score and Area Under the receiver operating characteristic Curve (AUC) as our evaluation metrics for performance comparison.

Pre-trained Model. We compare our model with different state-of-the-art methods on four standard manipulated datasets, NIST16 [18], CASIA (including CASIA 1.0 and CASIA 2.0) [7], COVER [23], and Columbia datasets [17]. According to the observation, except for CASIA, the other three standard datasets do not have enough samples for network training. Thus, we adopt a strategy that we use CASIA 2.0 to pre-train the model, then fine-tune the model on CASIA 1.0, NIST16, COVER, and Columbia.

Impact of Pixel Normalization. In order to emphasize the significance of the proposed Pixel Norm, when pre-training the model, we compare the performance of the model with Pixel Norm and without Pixel Norm. Under the same training conditions, the test results are shown in Fig. 4. Figure 4 intuitively shows the loss and F1 score of different scale patches and final segmentation with the network training. In contrast, the model with Pixel Norm can converge faster and the F1 score of different scale patches and final segmentation all improve.

Comparison Against Existing Methods. We compare our method with various state-of-the-art methods, including ELA [10], NOI [15], MFCN [21], J-LSTM [3], RGB-N [27], LSTM-ED-S [16], and ManTra-Net [24].

Table 1 shows the F1 score comparison results between our method and the other methods. Table 2 shows the AUC comparison results between our method and the other methods. It is worth noticing that both the F1 score and AUC of the CASIA dataset have two results. The value outside (.) is the result of model fine tuning and the value in (.) is the result of model non-fine tuning. From the results in Table 1 and Table 2, we can find that our model outperforms the other state-of-the-art methods on NIST16, CASIA, and Columbia. Especially

Table 1. F1 score comparison on four benchmarks. '-' denotes that the result is not available in the literature.

Method	NIST16	CASIA	Columbia	COVER
ELA [10]	0.236	0.214	0.470	0.222
NOI [15]	0.285	0.263	0.574	0.269
MFCN [21]	0.570	0.541	0.612	-
J-LSTM [3]	-	-	-	-
RGB-N [27]	0.722	0.408	0.697	**0.437**
LSTM-ED-S [16]	-	0.432	-	-
ManTra-Net [24]	-	-	-	-
Ours	**0.756**	**0.548**(0.469)	**0.902**	0.435

Table 2. AUC comparison on four benchmarks. '-' denotes that the result is not available in the literature.

Method	NIST16	CASIA	Columbia	COVER
ELA [10]	0.429	0.613	0.581	0.583
NOI [15]	0.487	0.612	0.546	0.587
MFCN [21]	-	-	-	-
J-LSTM [3]	0.764	-	-	0.614
RGB-N [27]	**0.937**	0.795	0.858	0.817
LSTM-ED-S [16]	0.857	0.814	-	-
ManTra-Net [24]	0.795	0.817	0.824	**0.819**
Ours	0.927	**0.858**(0.797)	**0.942**	0.813

on Columbia, the F1 score of our model increases by 0.205. One of the reasons is that images in Columbia are uncompressed and do not undergo any post-processing operations, which preserves the changes of statistical properties so well that the proposed model can accurately classify tampered patches.

For the poor performance on COVER, one possible explanation is that COVER only focuses on the copy-move operation which only causes the small change of statistical properties, and COVER is a relatively small dataset, not enough to fine-tune the model to adapt the new changes of statistical properties.

Qualitative Result. In Fig. 5, we provide the segmentation results of some examples produced by the proposed network. Different scale patch prediction masks resize to the same size as the original images. The prediction modules of different scale patches have completed the localization of the tampered region edges from coarse to fine. The fusion of semantic information and prediction masks also effectively eliminates the patch-effect.

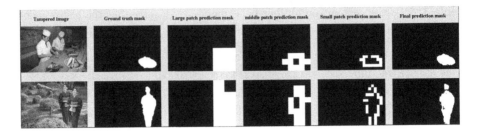

Fig. 5. Some of the segmentation examples.

5   Conclusions

In this paper, we propose an end-to-end model that can realize the coarse-to-fine localization of the tampered regions without any pre- and/or post-processing. The proposed model is sensitive to the statistical properties of the images and suitable for any manipulation types. The proposed Pixel Normalization proves that the classification ability of the model can be improved by appropriate constraints on the features. The experiments show the superiority of our model, which significantly outperforms the state-of-the-art works on four manipulated image benchmarks. For multi-task learning, the coefficient of each loss functions are predefined, but they may be not necessarily optimal. The dynamic adjustment strategy of the coefficients will be explored in the future.

References

1. Amerini, I., Uricchio, T., Ballan, L., Caldelli, R.: Localization of jpeg double compression through multi-domain convolutional neural networks. In: 2017 IEEE Conference on Computer Vision and Pattern Recognition Workshops (CVPRW), pp. 1865–1871. IEEE (2017)
2. Ba, J.L., Kiros, J.R., Hinton, G.E.: Layer normalization. arXiv preprint arXiv:1607.06450 (2016)
3. Bappy, J.H., Roy-Chowdhury, A.K., Bunk, J., Nataraj, L., Manjunath, B.: Exploiting spatial structure for localizing manipulated image regions. In: Proceedings of the IEEE International Conference on Computer Vision, pp. 4970–4979 (2017)
4. Bondi, L., Lameri, S., Güera, D., Bestagini, P., Delp, E.J., Tubaro, S.: Tampering detection and localization through clustering of camera-based CNN features. In: 2017 IEEE Conference on Computer Vision and Pattern Recognition Workshops (CVPRW), pp. 1855–1864. IEEE (2017)
5. Chang, I.C., Yu, J.C., Chang, C.C.: A forgery detection algorithm for exemplar-based inpainting images using multi-region relation. Image Vision Comput. **31**(1), 57–71 (2013)
6. Cozzolino, D., Verdoliva, L.: Noiseprint: a CNN-based camera model fingerprint. IEEE Trans. Inf. Forensics Secur. **15**, 144–159 (2019)
7. Dong, J., Wang, W., Tan, T.: Casia image tampering detection evaluation database. In: 2013 IEEE China Summit and International Conference on Signal and Information Processing, pp. 422–426. IEEE (2013)

8. Goljan, M., Fridrich, J.: CFA-aware features for steganalysis of color images. In: Media Watermarking, Security, and Forensics 2015, vol. 9409, p. 94090V. International Society for Optics and Photonics (2015)

9. Ioffe, S., Szegedy, C.: Batch normalization: Accelerating deep network training by reducing internal covariate shift. arXiv preprint arXiv:1502.03167 (2015)

10. Krawetz, N., Solutions, H.F.: A picture's worth. Hacker Factor Solutions **6**(2), 2 (2007)

11. Li, C., Ma, Q., Xiao, L., Li, M., Zhang, A.: Image splicing detection based on markov features in QDCT domain. Neurocomputing **228**, 29–36 (2017)

12. Li, G., Wu, Q., Tu, D., Sun, S.: A sorted neighborhood approach for detecting duplicated regions in image forgeries based on DWT and SVD. In: 2007 IEEE International Conference on Multimedia and Expo, pp. 1750–1753. IEEE (2007)

13. Lin, T.Y., Goyal, P., Girshick, R., He, K., Dollár, P.: Focal loss for dense object detection. In: Proceedings of the IEEE International Conference on Computer Vision, pp. 2980–2988 (2017)

14. Lyu, S., Pan, X., Zhang, X.: Exposing region splicing forgeries with blind local noise estimation. Int. J. Comput. Vision **110**(2), 202–221 (2014)

15. Mahdian, B., Saic, S.: Using noise inconsistencies for blind image forensics. Image Vision Comput. **27**(10), 1497–1503 (2009)

16. Mazaheri, G., Mithun, N.C., Bappy, J.H., Roy-Chowdhury, A.K.: A skip connection architecture for localization of image manipulations. In: Proceedings of the IEEE Conference on Computer Vision and Pattern Recognition Workshops, pp. 119–129 (2019)

17. Ng, T.T., Hsu, J., Chang, S.F.: Columbia image splicing detection evaluation dataset. Columbia Univ. CalPhotos Digit Libr, DVMM lab (2009)

18. Nimble, N.: Datasets (2016)

19. Pan, X., Zhang, X., Lyu, S.: Exposing image splicing with inconsistent local noise variances. In: 2012 IEEE International Conference on Computational Photography (ICCP), pp. 1–10. IEEE (2012)

20. Popescu, A.C., Farid, H.: Exposing digital forgeries in color filter array interpolated images. IEEE Trans. Signal Process. **53**(10), 3948–3959 (2005)

21. Salloum, R., Ren, Y., Kuo, C.C.J.: Image splicing localization using a multi-task fully convolutional network (MFCN). J. Visual Commun. Image Representation **51**, 201–209 (2018)

22. Wang, W., Dong, J., Tan, T.: Exploring DCT coefficient quantization effects for local tampering detection. IEEE Trans. Inf. Forensics Secur. **9**(10), 1653–1666 (2014)

23. Wen, B., Zhu, Y., Subramanian, R., Ng, T.T., Shen, X., Winkler, S.: Coverage-a novel database for copy-move forgery detection. In: 2016 IEEE International Conference on Image Processing (ICIP), pp. 161–165. IEEE (2016)

24. Wu, Y., AbdAlmageed, W., Natarajan, P.: Mantra-net: manipulation tracing network for detection and localization of image forgeries with anomalous features. In: Proceedings of the IEEE Conference on Computer Vision and Pattern Recognition, pp. 9543–9552 (2019)

25. Yang, Q., Peng, F., Li, J.T., Long, M.: Image tamper detection based on noise estimation and lacunarity texture. Multimed. Tools Appl. **75**(17), 10201–10211 (2016)

26. Zampoglou, M., Papadopoulos, S., Kompatsiaris, Y.: Large-scale evaluation of splicing localization algorithms for web images. Multimed. Tools Appl. **76**(4), 4801–4834 (2017)

27. Zhou, P., Han, X., Morariu, V.I., Davis, L.S.: Learning rich features for image manipulation detection. In: Proceedings of the IEEE Conference on Computer Vision and Pattern Recognition, pp. 1053–1061 (2018)

Security of AI-based Multimedia Applications

Efficient Generation of Speech Adversarial Examples with Generative Model

Donghua Wang, Rangding Wang[✉], Li Dong, Diqun Yan, and Yiming Ren

College of Information Science and Engineering, Ningbo University, Ningbo, Zhejiang, China
{wangrangding,dongli,yandiqun}@nbu.edu.cn

Abstract. Deep neural network-based keyword spotting (KWS) have embraced the tremendous success in smart speech assistant applications. However, the neural network-based KWS have been demonstrated susceptible to be attacked by the adversarial examples. The investigation of efficient adversarial generation would mitigate the security flaws of network-based KWS via adversarial training. In this paper, we propose to use the conditional generative adversarial network (CGAN) to efficiently generate speech adversarial examples. Specifically, we first present a target label embedding method to map the class-wise label into feature maps. Then, we utilize generative adversarial network for constructing the target speech adversarial examples with such feature maps. The target KWS classification network is then integrated with CGAN framework, where the classification error of the target network is back-propagated via gradient flow to guide the generator updating, but the target network itself is frozen. The proposed method is evaluated on a set of state-of-the-art deep learning-based KWS classification networks. The results validate the effectiveness of the generated adversarial examples. In addition, experimental results also demonstrate that the transferability of generated adversarial example among the different KWS classification networks.

Keywords: Speech adversarial examples · Conditional generative adversarial network · Keyword spotting (KWS)

1 Introduction

Nowadays, deep neural networks (DNNs) have become a critical backbone for speech recognition and verification systems. They have been extensively employed in many practical application, ranging from voice assistant on mobiles (i.e., Siri, Google Assistant, Alexa) to smart speaker on the intelligent home device (i.e., Apple Homepod, Amazon Echo). With increasing usage of those smart devices, the potential security risk aroused.

© Springer Nature Switzerland AG 2021
X. Zhao et al. (Eds.): IWDW 2020, LNCS 12617, pp. 251–264, 2021.
https://doi.org/10.1007/978-3-030-69449-4_19

The security risk of DNNs is introduced by the adversarial attack, a malicious attack approach that imposed carefully-crafted adversarial perturbation on the legitimate input of DNNs to cause misclassification. Recently, researchers attempted to investigate the possibility of adversarial attack in the speech domain. Some work [10,16] have demonstrated that legitimate voice inputs crafted by injecting small perturbations could mislead the DNN-based recognition systems. Several works also suggested that vulnerable of different speech-domain application to adversarial attack, including but not limited to speaker verification [3,14], keyword spotting (KWS) classification [1,5,26], and speech-to-text transcription [2,11,19].

Although existing works on the adversarial attack of the recognition systems have been reported, they are still existing challenges. More specifically, the state-of-the-art adversarial attack approaches in image domain or speech domain, making several unrealistic hypotheses on the time-cost setting, especially *having unlimited time for generating adversarial perturbation*. For image adversarial attack, the legitimate input images are typically static and constant. However, speech signals have different temporal characteristic: first, in real-world speech applications, the legitimate inputs are typically streaming voice input; therefore, the existing speech adversarial attacks, which depend on complicated iterative optimization approaches [2,27] and evolutionary algorithms [1,5], are too slow to realize those real-time speech processing systems attack; Second, it impossible for the adversary to generate adversarial speech perturbation during input-streaming phase due to the inherent sequential of speech signals, which the existing perturbation generation methods are based on the entire speech input.

To solve these issues, we attempt to use a CGAN-based method to generate speech adversarial perturbation. Specifically, we first present a target label embedding method to map the class-wise label information to feature maps, which used to guide the generator generate specific target perturbation; and then integrate the target network into the adopted CGAN framework. The goal of generator should meet two goals simultaneously: 1) deceiving the discriminator treating the adversarial example as the legitimate one, and 2) making the target network misclassified the adversarial example to specific target. The discriminator is to distinguish the constructed adversarial examples from the legitimate ones. The target network is fixed during training. Its classification loss is feedback via gradient flow to the generator, guiding the generator to adjust the perturbation towards the misclassification direction. Experimental results validate the attacking capability of proposed method on several widely-used KWS classification networks while maintaining imperceptual to human. The contributions of this work are summarized as follows

1) We present a target label embedding to map the target label to feature maps, which could enable us to use one trained generator to construct the adversarial example with the arbitrary specific target. By incorporating the target KWS classification network into our specially devised GAN framework, we successfully trained a generator that could generate the adversarial perturbation.

2) We conduct extensive experiments to demonstrate our generated adversarial examples could effectively fool the KWS classification network systems with high confidence, while retaining reasonably good perception quality. In addition, experimental results suggest that the transferability of the generated adversarial examples.

The rest of this work is organized as follows. We first briefly review related work in Sect. 2. Our proposed targeted speech adversarial example generation using CGAN is presented in Sect. 3, with a thorough discussion on the network architecture, loss function and training strategy. Experimental results are provided in Sect. 4, and finally we conclude this work in Sect. 5.

2   Related Work

In this section, we first review the recent generative adversarial network and its implementation in adversarial examples, and then survey the advance speech adversarial examples generation task.

2.1   Generative Adversarial Network

The generative adversarial network (GAN) was first proposed by Goodfellow *et al.* [6]. The most significant idea of GAN is to establish a game between two networks, i.e., the generator and discriminator network. After sufficient training, the generator obtains the ability to generate the samples whose distribution resembles to the training data. Mirza *et al.* [15] introduced the conditional version of GAN (CGAN). The critical point of CGAN is introducing the label information into generator, which could generate samples conditioned on label. CGAN has achieved tremendous success in many computer vision tasks such as conditional image generation [17], image-to-image translation [9], and text-to-image synthesis [20]. Compared with the extensive study of GAN in the image domain, there are far less works in the speech domain. Pascual *et al.* [18] devised a generative adversarial network for speech enhanced task, which could remove noise from the raw speech input and obtain more clear speech clip. Recently, GAN is treated as a kind of adversarial examples generation method. For examples, Hu *et al.* [8] proposed to use GAN to generate adversarial malware examples, which could bypass the detection systems. Xiao [24] *et al.* used GAN to generate realistic adversarial image examples, which could learn the latent adversarial representation that deceives the victim model. Wang *et al.* [22] proposed to use GAN to generate adversarial examples, but they need to train a model for each target, which is lower-efficient in adversarial examples generation.

2.2   Speech Adversarial Example

Recently, adversarial examples attack in the speech domain has arouse attention of researchers. Carlini *et al.* [2] utilized the fast gradient sign method (FGSM) to generate speech adversarial perturbation. The results have illustrated that the generated adversarial examples can successfully degrade the performance of speech recognition network. Alzantot et al. [1] proposed to utilize the genetic algorithm to search adversarial examples from a set of candidate population. The sought adversarial example was successfully misclassified to the assigned class. Inspired by [1], Du et al. [5] proposed a particle swarm optimization-based method to seek speech adversarial examples. These two methods involve solving a complicated optimization problem with several heuristic tricks. However, the computational cost of the above optimization-based methods is quite heavy. Recently, Xie et al. [26] proposed to utilize U-Net-based model with target class embedding to generate adversarial perturbation. Unlike the aforementioned approaches, our proposed framework implants the target network to a conditional generative adversarial network, and uses the gradient information feedback by the target network to guide the training of the generator. Once completing the training, the generator can efficiently generate perturbation for a given speech and target label.

3   Adversarial Examples Generation with GAN

In this section, we first propose the problem formulation and target label embedding. Then the network architecture of our proposed method is presented. Following that, the loss function is thoroughly discussed, and finally the training strategy is provided.

3.1   Problem Formulation

Let $\mathcal{X}$ be the original speech waveform space, and $\mathcal{Y}$ be the ground-truth label space. A well-trained keyword spotting (KWS) classification network $f(\cdot)$, which is expected to receive a speech sample clip $x \in \mathcal{X}$, and output its corresponding category $y \in \mathcal{Y}$, i.e., $f(x) = y$. Given such KWS classification network $f(\cdot)$, an input speech sample x, and a pre-specified target category $y_t \neq y$, the goal of our targeted speech adversarial example generation scheme have two-folds. First, we would like to generate a speech adversarial example x' that could effectively deceive the given KWS classification network, i.e.,

$$f(x') = y_t, \quad \text{where } y_t \neq y. \tag{1}$$

Second, the generated adversarial example x' should be similar to the original sample x in terms of human auditory perception. That is, the common listener cannot tell the different between x and x' when listening to them.

More formally, to realize the aforementioned two goals, we aim to find a tiny additive perturbation δ such that $x' = x + \delta$. Under this formulation, the adversarial example generation algorithm can be expressed as the following optimization problem

$$\delta^* = \arg\min_{\delta} \ell(f(x + \delta), y_t) \tag{2}$$

$$\text{s.t.: } y_t \neq y, \text{ and } \|\delta\| < \epsilon. \tag{3}$$

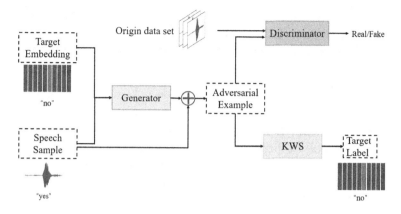

Fig. 1. The overview of the proposed framework.

where $\ell(\cdot)$ is the loss function to evaluate the classification accuracy, if and only if $f(x') = y_t$ it reaches the minima. Here $\|\cdot\|$ is the similarity metrics ℓ_2, and ϵ denotes the maximum perturbation magnitude.

3.2 Target Label Embedding

Inspired by the conditional generative adversarial network [4,12], we design a target label embedding. In order to motivate the generator to learn the target class information, we propose to using the k-class label embedding method. That method is used to ensure that a single trained model could be re-used for attacking against various of target classes instead of single-specific design. Formally, the embedding feature map of class-aware target label can be expressed as following

$$\xi = \{E_1, E_2, ..., E_k\}. \tag{4}$$

where ξ is the embedded feature map, and E_k represents the embedding feature map of class k. ξ is designed to be trainable, each of them corresponds to one specific target class. Given an arbitrary target class y_t, the target label feature map E_t is extracted through target label embedding. The feature map is concatenated with the speech signal to form the input of model.

3.3 Framework Overview

Figure 1 illustrates the overview architecture of the proposed framework. In general, our framework is mainly composed of three components: a generator $\mathcal{G}$, a discriminator $\mathcal{D}$ and the target KWS classification network f. This is a competitive learning among three parties: generator $\mathcal{G}$, discriminator $\mathcal{D}$ and KWS classification network f, where the KWS part is fixed. The crucial parts of our method is to utilize the target label embedding and the conditional generative adversarial network (CGAN) to generate adversarial example. The goal of the

GAN components is to learn the latent representation to deceive the target model, under small perturbation constraint. Mathematically, the original speech x and specific target label embedding ξ is fed into generator and its output $\mathcal{G}(x, \xi)$ as the adversarial perturbation, which can be formulated as

$$x' = x + \mathcal{G}(x, \xi). \tag{5}$$

Then, x' is fed into the discriminator $\mathcal{D}$ and the target model f, respectively. The function of discriminator $\mathcal{D}$ is to distinguish the generated adversarial examples from the benign ones, forcing the generator $\mathcal{G}$ to produce the perturbation that is indistinguishable to the discriminator. The target classifier f is involved to guide the generator to craft generated examples that would be misclassified to a pre-specified target label.

3.4 Loss Function

To train a generator that misleads the KWS classification network to misclassify adversarial examples to a specified target, the loss is a critical point to ensure the training process. In general, there are two networks, i.e., generator $\mathcal{G}(\cdot)$ and discriminator $\mathcal{D}(\cdot)$, to be trained. Each network shall be trained with a well designed loss function. More specifically, for the generator, we define its loss function as

$$\mathcal{L}_\mathcal{G} = \mathcal{L}_{adv} + \alpha \mathcal{L}_f + \beta \mathcal{L}_2, \tag{6}$$

where $\mathcal{L}_{adv}$ is the loss to denote the fooling capability of generator on the discriminator; $\mathcal{L}_f$ is the adversarial loss, representing the attacking ability of generator on the target classifier; The last term $\mathcal{L}_2$ is ℓ_2-norm loss of the adversarial perturbation, which is adopted here to expect an acceptable auditory quality of speech adversarial example. The parameters α and β are the weights to balance the importance among the three loss terms. We will give the more detailed explanation on each term in (6) as follows

Adversarial Loss $\mathcal{L}_{adv}$: To encourage the generator to seek an appropriate perturbation that misleads the classification of discriminator, the generator receives the feedback of discriminator to enhance its fooling ability consistently. To this end, the adversarial loss is designed as follows

$$\mathcal{L}_{adv} = \mathbb{E}_x[log(1 - D(x'))], \tag{7}$$

where $\mathbb{E}_x[\cdot]$ denotes the exception operator over the training data. It can be seen that minimizing $\mathcal{L}_{adv}$ is equivalent to maximizing the classification probability $D(x')$ towards to 1. This essentially guides the constructed adversarial example to mimic the legitimate speech data distribution.

Target Classifier Loss $\mathcal{L}_f$: One primary goal of the generator is to deceive the target KWS classification network. Then the classifier loss $\mathcal{L}_f$ of target network shall measure the distance between the targeted label with the classifier prediction of the adversarial example. Mathematically, we have

$$\mathcal{L}_f = \mathbb{E}_x[\ell_{ce}(f(x', y_t))], \tag{8}$$

where y_t is the specific target label of an adversary, x' is the constructed adversarial examples, and $\ell_{ce}(\cdot)$ denotes the cross-entropy loss function that commonly used in multiple classification tasks. As one can see, the loss $\mathcal{L}_f$ encourages the constructed adversarial speech examples are classified to the target label y_t.

Perturbation Constrain Loss $\mathcal{L}_2$: The goal of $\mathcal{L}_2$ is to bound the magnitude of perturbation, that is to say it can ensure the auditory quality of constructed adversarial examples. To make adversarial examples as similar as the original one, we shall ℓ_2-norm to bound the magnitude of the adversarial perturbation, i.e.,

$$\mathcal{L}_2 = \|x' - x\|_2. \tag{9}$$

This loss function could control the total magnitude of the perturbation, avoiding the perturbation exceed the normal range. That is, the perturbation magnitude should keep smaller when training the generator.

For the discriminator, its goal is to distinguish the generated adversarial examples(fake) from the legitimate ones(true). We adopt the following loss function

$$\mathcal{L}_\mathcal{D} = \frac{1}{2}\left(\mathbb{E}_x\left[\log(1 - \mathcal{D}(x))\right] + \mathbb{E}_x\left[\log(\mathcal{D}(x'))\right]\right), \tag{10}$$

where $\mathcal{D}(x') \in [0,1]$ denotes the probability that the adversarial example is classified as true by the discriminator $\mathcal{D}$. Note that this loss function for discriminator is similar to the one used in LSGAN [13], in which could make the GAN training procedure more stable.

Finally, by combining the loss function for generator ($\mathcal{L}_G$) and discriminator ($\mathcal{L}_D$), we can acquire the generator by solving the following min-max optimization problem

$$\mathcal{G}^\star = \arg\min_\mathcal{G}\max_\mathcal{D}\left(\mathcal{L}_G + \mathcal{L}_D\right), \tag{11}$$

where $\mathcal{G}^\star$ is a well-trained generator. Once training completed, one can efficiently generate the adversarial for the given input and arbitrary target with $\mathcal{G}^\star$. However, training (11) is not a trivial task. In the next section, we discuss the training strategy that deliberately designed for the targeted adversarial examples generation task.

3.5   Training Strategy

During the training phase, the generator and discriminator are trained in an alternating way, i.e., when training the generator, the discriminator is fixed and vice versa. It is also worth pointing out that the network parameters of the target network are fixed during the entire training stage. That is, the KWS classification network prediction errors can be backpropagated via gradients to the generator $\mathcal{G}$, but the weights of the target network itself is not updated.

4   Experimental Result

In this section, we first introduce the dataset and experimental setting that are used throughout the experiments. Then, a number of state-of-the-art KWS classification networks is implemented, and retreat as our victim models. Following that, the perceptual quality of the generated adversarial example is evaluated. Finally, the robustness of the generated adversarial examples is discussed.

4.1   Experimental Settings

Dataset. Our experiments are conducted on the following datasets: the Google single-word speech command dataset `SpeechCmd` [23]. The `SpeechCmd` consists of 65000 audio files of 30 single words. Each word file is a one-second speech clip of a single word, the following ten classes are selected as our dataset: "yes", "no", "up", "down", "left", "right", "on", "off", "stop", and "go". We split it into a training set, a validation set and a test set in the ratio 80%: 10%: 10%. Experimental results are all based on the test set. As practiced in previous work [18], the input of model is preprocessing with normalized into $[-1, 1]$. The α is set to 1, β is set to 100.

Evaluation Metric. The mainly performance metric of adversarial attack against to KWS classification network is the attack success rate (success_rate) of the generated adversarial examples, which can be expressed as

$$success\_rate = \frac{\#\{\text{misclassified samples}\}}{\#\{\text{test samples}\}}, \tag{12}$$

where $\#\{\cdot\}$ returns the cardinality of the input set.

SNR is commonly used to evaluate the quality of generated adversarial speech in previous works [5,27]. Therefore, we use SNR to evaluate the audio quality of generated adversarial examples. SNR is calculated as follows

$$SNR(x') = 10 \cdot \log_{10} \frac{P_x}{P_\delta}, \tag{13}$$

where x', x are the adversarial example and the geniue speech, respectively; δ is the generated perturbation noise. P_x and P_δ denote the energies of the intrinsic speech signal and the perturbation noise, respectively. Specifically, a large SNR value indicates a small noise scale.

4.2   Attacking to the Network-Based KWS Classifiers

The attacking is assumed implemented under the white-box scenario, where an adversary can utilize the gradient of the target network for attacks. More specifically, in the forward propagation phase of the entire training procedure, for each batch of input speech data x, we random select batch target class y_t,

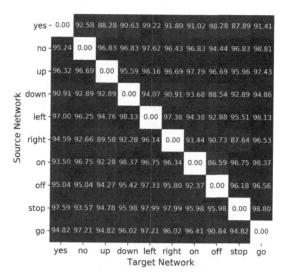

Fig. 2. The confusion matrix of the success_rate for the ResNet18-based KWS classification network [7] on the SpeechCmd dataset. Our targeted attack enforces a given speech that originally belongs to the source label misclassified as the target label. The diagonal entries are all set as zeros because they indicate the source equals target, which is a correct classification rather than an intended attack.

and fetch the corresponding embedding feature map ξ. The selected feature maps are concatenated to the raw speech signal as input feed to generator $\mathcal{G}$ to form perturbation $\mathcal{G}_t(x, \xi)$. Then, the constructed adversarial examples are fed into the target network to compute the loss between the output and the batch of the target label. The network parameters of the generator are then updated with the gradient information that back-propagated from the loss. Finally, we can obtain a well-trained model to realized the targeted attack on a KWS classification network.

The attacking experiments are conducted on four state-of-the-art KWS classification networks: WideResNet [28], VGG19 [21], ResNet18 [7], ResNeXt [25]. These are well known for their good classification performance in the Tensorflow Speech Recognition Challenge[1]. Therefore, we modify them to adapt the spectrogram input.

In attacking specific a target network, we generate the adversarial examples for ten targets label with the trained generator over test dataset. Figure 2 illustrates the attack success rate results for the ResNet18 [7], which is presented in the form of confusion matrix. Note that, the diagonal values of the confusion matrix are all intentionally set are zero. This is because diagonal entry indicates the source equals target, which is a correct classification rather than an intended attack. As shown in Fig. 2 the success_rate of our method against the ResNet18 network generally larger than 94%(the average success_rate is 94.81%). In other

---

[1] https://www.kaggle.com/c/tensorflow-speech-recognition-challenge.

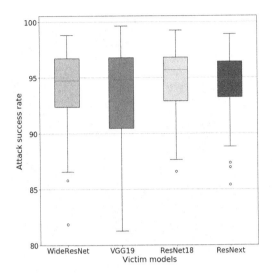

Fig. 3. The bot-plot of the success_rate for the four KWS classification networks on the SpeechCmd dataset.

words, the average classification accuracy of the target KWS network on generated adversarial examples is about 5.19%, which means that our generated adversarial examples could nearly paralyze the ResNet18-based KWS classifier. Take the target attack "yes" → "left" as an example. Our generated adversarial examples which sound like "yes" could be misclassified as the opposite "left" by ResNet18-based KWS classifier of high confidence 99.22%. This could cause high security risks for some critical KWS classifiers, e.g., the KWS equipped in the smartphone. With more carefully examination, one can notice that, on one hand, the success_rate of the target "left" and "go" are all exceeding 96% (the average success_rate's for "no", "up", are 97.17%, 96.77%, respectively). This implies that these particular words are more prone to be as the targets when attacking the ResNet18-based KWS classifier. On the other hand, when the source words are "no", "up", "left" and "stop", the average attack success_rate are 96.65%, 96.81%, 96.05% and 96.52%, all exceeding 96%. This phenomenon suggests that these words would be much easier to be deceived, which many guide the adversary to exploit such vulnerable words when implementing an attack. Similar observations can also be made in another victim model. In short summary, the success_rate results verify that our method can effectively attack the target network-based KWS classifier.

To better illustrate the results, we further present a boxplot to show the attack success rate, against the different model. The comparison results are shown in Fig. 3. From the boxplot in Fig. 3, on the one hand, we can observe that the median of success_rate of WideResNet, VGG19, ResNet18 and ResNeXt are all exceeding 94% (94.74%, 94.79%, 95.71% and 94.85%, respectively). In the other hand, the outlier points are few with the value are all exceed 80%. The results suggest that our method could successful attack different KWS classifier.

Table 1. Comparison the proposed method with Alzantot *et al.* [1] and SirenAttack [5] on WideResNet [28], VGG19 [21], ResNet18 [7], ResNeXt [25]

Model/Method	Accuracy	Alzantot [1]			SirenAttack [5]			Proposed		
		success_rate	SNR(dB)	Time(s)	success_rate	SNR(dB)	Time(s)	success_rate	SNR(dB)	Time(s)
WideResNet	94.49%	83.27%	14.51	37.28	89.25%	17.57	368.29	**94.04%**	**18.77**	**0.008**
VGG19	91.35%	81.39%	13.53	34.81	88.10%	18.22	332.26	**93.01%**	**18.54**	**0.006**
ResNet18	92.13%	80.25%	13.72	44.35	87.35%	15.87	340.31	**94.81%**	**18.25**	**0.008**
ResNeXt	93.57%	86.14%	14.85	33.87	90.05%	17.03	317.92	**94.52%**	**18.85**	**0.007**

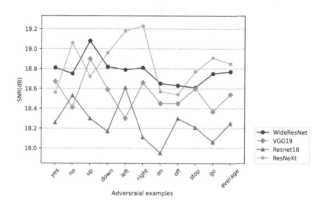

Fig. 4. The comparison of the SNR for four KWS classification network on the SpeechCmd dataset.

Furthermore, we compare the proposed method with two recently state-of-the-art works, Alzantot *et al.* [1] and SireAttack [5], against four target models on SpeechCmd. The comparison results are listed in Table 1, from which one can observe that our method generally achieves superior performance. For instance, when attacking ResNet18, the average success_rate for [1] is 80.25% with average SNR 13.72 dB, and the average time for generating an adversarial speech clip is 44.35 s; The average success_rate for [5] is 87.35% with average SNR 15.87 dB, and the average time for generating an adversarial speech clip is 340.31 s. For comparison, our proposed method yields attack success_rate with 94.81% with average SNR 18.25 dB, but only takes 0.0008 s for generating an adversarial example on average.

4.3 Quality Evaluation of the Generated Adversarial Examples

We compute the average SNR values between the original sample with the generated adversarial example for each target. The results are shown in Fig. 4. Specifically, the highest average SNR is 18.85 dB for ResNeXt. As illustrated in Fig. 4, when the target model is ResNeXt, the SNR value of generated adversarial examples is the highest. Further analysis with Table 1, the adversarial examples with lower SNR value could achieve higher success_rate and well transferability, for example, adversarial examples generated by generator trained on ResNet18 could achieve higher success_rate (94.81%) with the lowest SNR among four victim models.

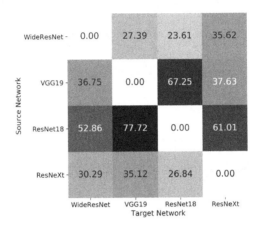

Fig. 5. Confusion matrix of the attack success rate of transferability among different victim models.

4.4   Transferability of Generated Adversarial Example

Finally, we assess whether the generated adversarial examples against one specific network can fool another networks. We use the adversarial examples generated in Sect. 4.2. For example, we choose the ResNet18 network as our target network, let another three networks WideResNet, ResNeXt, VGG19 as the unknown networks. Adversarial examples generated by ResNet18 are used to attack another three networks. The comprehensive results are listed in Fig. 5. As shown in Fig. 5, we can achieve success_rate with the result that min 23.61%, max 77.72%, average 40.19%. Take the "ResNet18" → "VGG19" network as example. Adversarial examples generated by the trained generator on ResNet18 to attack the "VGG19" with 77.72% average success_rate, which means the 77.72% of adversarial examples generated by ResNet18 can deceive the VGG19. Furthermore, adversarial example with the highest capacity of transferability is generated by the generator trained with ResNet18, for which these adversarial examples have lower SNR value that indicates more perturbation is introduced. Moreover, from Fig. 5 we can see, the average success_rate of adversarial examples generated with the generator trained with WideResNet is 28.05%, which means these adversarial examples have the worse transferability. In contrast, the WideResNet is most robust against adversarial examples generated by the generator trained with other models.

5 Conclusion

In this work, we utilize the conditional adversarial generative network to generate speech adversarial examples. More specifically, we first presented a target label embedding that could map the target label to feature map. Combined with generative adversarial networks, one could generate speech adversarial examples with the specific target label, which could successfully paralyze network-based KWS classification networks. Our proposed method integrates the target network into generative adversarial network framework. Once completing the training, the well-trained generator could efficiently generate adversarial samples for the arbitrary specific target. Experiments demonstrate the transferability of the generated adversarial examples among the different state-of-the-art classification networks.

Acknowledgements. This work is supported by National Natural Science Foundation of China (Grant No. U1736215, 61672302, 61901237), the Zhejiang Natural Science Foundation (Grant No. LY20F020010, LY17F020010), the Ningbo Natural Science Foundation (Grant No. 2019A610103, 202003N4089) and K.C. Wong Magna Fund in Ningbo University.

References

1. Alzantot, M., Balaji, B., Srivastava, M.: Did you hear that? Adversarial examples against automatic speech recognition. arXiv preprint arXiv:1801.00554 (2018)
2. Carlini, N., Wagner, D.: Audio adversarial examples: targeted attacks on speech-to-text. In: 2018 IEEE Security and Privacy Workshops (SPW), pp. 1–7. IEEE (2018)
3. Chen, G., et al.: Who is real bob? Adversarial attacks on speaker recognition systems. arXiv preprint arXiv:1911.01840 (2019)
4. Chen, X., Duan, Y., Houthooft, R., Schulman, J., Sutskever, I., Abbeel, P.: Info-GAN: interpretable representation learning by information maximizing generative adversarial nets. In: Advances in Neural Information Processing Systems (NeurIPS), pp. 2172–2180 (2016)
5. Du, T., Ji, S., Li, J., Gu, Q., Wang, T., Beyah, R.: SirenAttack: generating adversarial audio for end-to-end acoustic systems. arXiv preprint arXiv:1901.07846 2(1) (2019)
6. Goodfellow, I., et al.: Generative adversarial nets. In: Advances in Neural Information Processing Systems (NeurIPS), pp. 2672–2680 (2014)
7. He, K., Zhang, X., Ren, S., Sun, J.: Deep residual learning for image recognition. In: Proceedings of the IEEE Conference on Computer Vision and Pattern Recognition (CVPR), pp. 770–778 (2016)
8. Hu, W., Tan, Y.: Generating adversarial malware examples for black-box attacks based on GAN. arXiv preprint arXiv:1702.05983 (2017)
9. Isola, P., Zhu, J.Y., Zhou, T., Efros, A.A.: Image-to-image translation with conditional adversarial networks. In: Proceedings of the IEEE Conference on Computer Vision and Pattern Recognition (CVPR), pp. 1125–1134 (2017)
10. Kereliuk, C., Sturm, B.L., Larsen, J.: Deep learning and music adversaries. IEEE Trans. Multimed. **17**, 2059–2071 (2015)

11. Kwon, H.W., Kwon, H., Yoon, H., Choi, D.: Selective audio adversarial example in evasion attack on speech recognition system. IEEE Trans. Inf. Forensics Secur. **15**, 526–538 (2020)

12. Lee, C.Y., Toffy, A., Jung, G.J., Han, W.J.: Conditional WaveGAN. arXiv preprint arXiv:1809.10636 (2018)

13. Mao, X., Li, Q., Xie, H., Lau, R.Y., Wang, Z., Paul Smolley, S.: Least squares generative adversarial networks. In: Proceedings of the IEEE International Conference on Computer Vision, pp. 2794–2802 (2017)

14. Meng, Z., Zhao, Y., Li, J., Gong, Y.: Adversarial speaker verification. In: 2019 IEEE International Conference on Acoustics, Speech and Signal Processing (ICASSP), ICASSP 2019, pp. 6216–6220. IEEE (2019)

15. Mirza, M., Osindero, S.: Conditional generative adversarial nets. arXiv preprint arXiv:1411.1784 (2014)

16. Neekhara, P., Hussain, S., Pandey, P., Dubnov, S., McAuley, J., Koushanfar, F.: Universal adversarial perturbations for speech recognition systems. arXiv preprint arXiv:1905.03828 (2019)

17. Odena, A., Olah, C., Shlens, J.: Conditional image synthesis with auxiliary classifier GANs. In: International Conference on Machine Learning (ICML), pp. 2642–2651 (2017)

18. Pascual, S., Bonafonte, A., Serra, J.: SEGAN: speech enhancement generative adversarial network. arXiv preprint arXiv:1703.09452 (2017)

19. Qin, Y., Carlini, N., Cottrell, G., Goodfellow, I., Raffel, C.: Imperceptible, robust, and targeted adversarial examples for automatic speech recognition. In: International Conference on Machine Learning (ICML), pp. 5231–5240 (2019)

20. Reed, S., Akata, Z., Yan, X., Logeswaran, L., Schiele, B., Lee, H.: Generative adversarial text to image synthesis. arXiv preprint arXiv:1605.05396 (2016)

21. Simonyan, K., Zisserman, A.: Very deep convolutional networks for large-scale image recognition. In: International Conference on Learning Representations (ICLR) (2015)

22. Wang, D., Dong, L., Wang, R., Yan, D., Wang, J.: Targeted speech adversarial example generation with generative adversarial network. IEEE Access **8**, 124503–124513 (2020)

23. Warden, P.: Speech commands: a public dataset for single-word speech recognition. Dataset available from http://download.tensorflow.org/data/speech_commands_v0 1 (2017)

24. Xiao, C., Li, B., Zhu, J.Y., He, W., Liu, M., Song, D.X.: Generating adversarial examples with adversarial networks. arXiv abs/1801.02610 (2018)

25. Xie, S., Girshick, R., Dollár, P., Tu, Z., He, K.: Aggregated residual transformations for deep neural networks. In: Proceedings of the IEEE Conference on Computer Vision and Pattern Recognition (CVPR), pp. 1492–1500 (2017)

26. Xie, Y., Li, Z., Shi, C., Liu, J., Chen, Y., Yuan, B.: Enabling fast and universal audio adversarial attack using generative model. arXiv preprint arXiv:2004.12261 (2020)

27. Yuan, X., et al.: CommanderSong: a systematic approach for practical adversarial voice recognition. In: 27th USENIX Security Symposium (USENIX Security 2018), pp. 49–64 (2018)

28. Zagoruyko, S., Komodakis, N.: Wide residual networks. arXiv preprint arXiv: 1605.07146 (2016)

Lightweight DCT-Like Domain Forensics Model for Adversarial Example

Junjie Zhao[1] and Jinwei Wang[1,2](✉)

[1] Nanjing University of Information Science and Technology, Jiangsu 210044, China
wjwei_2004@163.com
[2] State Key Laboratory of Mathematical Engineering and Advanced Computing,
Henan 450001, China

Abstract. Since the emergence of adversarial examples brings great security threat to deep neural network which is widely used in various fields, their forensics become very important. In this paper, a lightweight model for the forensics of adversarial example based on DCT-like domain is proposed. The DCT-like layer realizes the block conversion of data from the spatial domain to the frequency domain. Together with the color space transformation layer and the residual layer, the DCT-like layer realizes the simulation of JPEG quantization error. The feature statistical layer is used to obtain the statistical feature values of the feature map output by the frequency-division convolution, and at the same time, it also contains learnable hyperparameters. The group BN strategy ensures the effectiveness of the DCT-like layer and the feature statistical layer and promotes the accuracy of forensics. Experiments show that the proposed model not only reaches the highest accuracy we know, but also it only needs to train for one epoch to get a high-performance.

Keywords: Deep neural network · Adversarial example · Forensics

1 Introduction

Due to the good performance of the convolutional neural networks (CNNs) across fields, it is widely used in various multimedia fields, including image recognition [8], semantic segmentation [5], video processing [1], and so on. However, the discovery of adversarial examples [20] exposes the vulnerability of neural networks. As long as a small disturbance that human vision cannot detect is added

This work was jointly supported by the National Natural Science Foundation of China (Grant No. 61772281, 61702235, 62072250, U1636117, and U1636219), in part by the National Key R&D Program of China(Grant No. 2016YFB0801303 and 2016QY01W0105), in part by the plan for Scientific Talent of Henan Province (Grant No.2018JR0018), in part by Postgraduate Research & Practice Innovation Program of Jiangsu Province (Grant No. KYCX20_0974) and the Priority Academic Program Development of Jiangsu Higher Education Institutions (PAPD) fund.

X. Zhao et al. (Eds.): IWDW 2020, LNCS 12617, pp. 265–279, 2021.
https://doi.org/10.1007/978-3-030-69449-4_20

to the sample to be tested, the convolutional neural network can draw completely wrong conclusions. With the iterative upgrade of the adversarial methods [10,15], a large number of adversarial examples continue to emerge. Since traditional forensics methods are powerless in the face of massive data, a rapid forensics scheme against adversarial examples needs urgently to be proposed.

At present, the forensics methods of adversarial examples are divided into traditional manual feature extraction methods and CNN-based detection methods. Most of the existing detection schemes are based on the former, which is possibly related to the risk of a secondary attack of the detection models based on CNNs. Grosse et al. [7] found that the difference of the maximum value, average value, and energy distribution between the examples can be used to detect adversarial examples. Li and Li [11] proposed a detection method that uses SVM to classify classifier-level features. This method directly obtains the output feature maps of each convolutional layer of the original classifier and sends the statistical features of them to the SVM for classification. Liu et al. [13] used a set of steganographic analysis methods and proposed an enhanced version of SRM (ERSM) based on the spatial domain rich model (SRM) [4] to achieve high accuracy of the detection of adversarial examples. Their work reveals similarities between the forensics of adversarial examples and steganalysis. In these methods, the process of classification is separated from the process of feature extraction, and end-to-end forensics cannot be realized. Besides, the statistical values of examples play a fundamental role in all of these methods. This has a great similarity with other image forensics fields, such as JPEG recompression forensics [22] and photograph (PG) and computer-generated (CG) images forensics [16,21].

Carrara et al. [3] proposed a typical detection scheme based on the CNN model. They extracted the neuron output of fully connected layer of the original classifier, and a long short-term memory (LSTM) [19] model is used to detect adversarial examples. This method relies on the results of the inner layer of the original model and cannot achieve end-to-end forensics, too.

Our experiments show that when the errors generated during the JPEG compression process, including conversion errors, quantization errors, truncation errors, and rounding errors [22], are sent to the neural network for training, a good forensic result can be obtained. Among these errors, the values of quantization errors are usually the largest and occupy the most important part of the effective features. Also, the effectiveness of steganalysis methods proves the feasibility of frequency domain forensics. Inspired by these phenomena, we designed a lightweight adversarial examples forensics model based on DCT-like domain, which mainly contains the following innovations:

- The convolution layer is used to simulate the block Discrete Cosine Transformation (DCT) to optimize the transformation coefficient along with the gradient update and transform the image from the spatial domain to a new DCT-like domain, which is more suitable for forensics. At the same time, the transformation process from RBG color space to YCbCr space is sim-

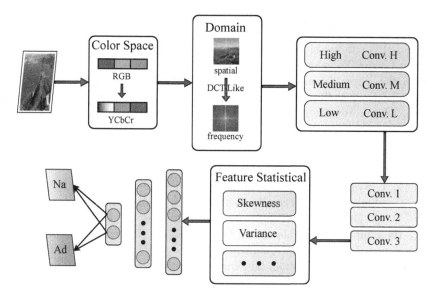

Fig. 1. The Proposed Model. The image in the upper left corner denotes the example to be determined. The final output "Na" and "Ad" represent natural examples and adversarial examples.

ulated, and the Chroma Green (Cg) channel is extended to obtain better performance.

- A feature statistical layer is proposed to directly obtain the maximum, variance, skewness, and other statistical values of the feature maps output from convolutional layers, and these values become effective for forensics under the effect of group regularization. According to the type of feature, the parameters are adopted to give the feature statistical layer learning ability.
- The Remainder layer is constructed to enlarge the difference between sample classes. The combination of the residual layer, DCT-like layer, and IDCT-like layer is taken to simulate the quantization error in the JPEG compression process, and the amplification of feature difference expands the room for accuracy improvement.

The rest of this paper is as follows: In Sect. 2, the proposed model is elaborated, including DCT-like layer, Remainder layer, Color Space Transformation, Feature Statistical layer and Group BN Strategy. In Sect. 3, experiments are conducted. And in Sect. 4, we conclude.

2   The Proposed Method

The overall structure of the forensics model proposed in this paper is shown in Fig. 1. It includes color space conversion layer, DCT-like layer, remainder layer, frequency division convolutional layer, feature statistical layer, full connection layer, etc.

2.1 Quantization Error and DCNN

During JPEG compression, the image is Transformed in 8×8 small blocks. The Discrete Cosine Transformation (DCT) can be expressed by matrix multiplication. In this paper, $\times$ means matrix multiplication, and $\cdot$ means the multiplication of corresponding elements in the matrixes. Let $\boldsymbol{X}$ represent the input sample, $\boldsymbol{M_{DCT}}$ represent the DCT matrix, and the transformation process can be represented by Eq. 1.

$$\boldsymbol{X_{DCT}} = Flatten(\boldsymbol{X}) \times \boldsymbol{M_{DCT}} \tag{1}$$

In Eq. 1, $Flatten()$ means to divide $\boldsymbol{X}$ into small blocks of size 8×8 and flatten the elements in each small block into one-dimensional form. Let $SDCT()$ denote Block DCT, $SIDCT()$ denote Block Inverse DCT (IDCT), and $\boldsymbol{M_Q}$ denote the quantization table of JPEG compression. Let $round()$ represent the rounding function, then the quantization error $\boldsymbol{ERR_Q}$ generated during the JPEG compression process can be expressed by Eq. 2.

$$\boldsymbol{ERR_Q} = \boldsymbol{X} - SIDCT(round(\frac{SDCT(\boldsymbol{X})}{\boldsymbol{M_Q}}) \cdot \boldsymbol{M_Q}) \tag{2}$$

$$floor(SDCT(\boldsymbol{X})/\boldsymbol{M_Q}) = SDCT(\boldsymbol{X})/\boldsymbol{M_Q} - SDCT(\boldsymbol{X})\%\boldsymbol{M_Q} \tag{3}$$

$$\begin{aligned} \boldsymbol{ERR_Q} &= \boldsymbol{X} - SIDCT((SDCT(\boldsymbol{X})/\boldsymbol{M_Q} - SDCT(\boldsymbol{X})\%\boldsymbol{M_Q}) \cdot \boldsymbol{M_Q}) \\ &= SIDCT(SDCT(\boldsymbol{X}) - (SDCT(\boldsymbol{X})/\boldsymbol{M_Q} - SDCT(\boldsymbol{X})\%\boldsymbol{M_Q}) \cdot \boldsymbol{M_Q} \\ &= SIDCT(SDCT(\boldsymbol{X}) - SDCT(\boldsymbol{X}) + SDCT(\boldsymbol{X})\%\boldsymbol{M_Q} \cdot \boldsymbol{M_Q}) \\ &= SIDCT(SDCT(\boldsymbol{X})\%\boldsymbol{M_Q} \cdot \boldsymbol{M_Q}) \end{aligned}$$

$$\tag{4}$$

$$\boldsymbol{ERR_Q} = ConvIDCT(Remainder(ConvDCT(\boldsymbol{X})/\boldsymbol{M_Q}) \cdot \boldsymbol{M_Q}) \tag{5}$$

In this paper, $round()$ in Eq. 2 is rounding down operation. Let $floor()$ represent this operation, and $\%$ denotes the remainder symbol. Then Eq. 3 is gotten. Replace the $round(SDCT(\boldsymbol{X})/\boldsymbol{M_Q})$ in Eq. 2 with the right half of Eq. 3, then Eq. 4 is gotten.

Assuming that convolutional layer ConvIDCT and ConvDCT are used to simulate $SIDCT()$ and $SDCT()$ respectively, and the remainder layer, $Remainder()$, is used to simulate $\%$. Substitute them into Eq. 4, then Eq. 5 could be obtained.

Equation 5 shows that by constructing ConvIDCT, ConvDCT, and Remainder in the DCNN, the quantization error can be easily obtained using the neural network.

2.2 DCT-Like Layer

In the DCT process of JPEG-compression, the input example $\boldsymbol{X}$ is divided into small blocks with a size of 8×8, and then each block is flattened into a one-dimensional vector. Using $Width$ and $Height$ to express the original image width and height. After partitioning, $\boldsymbol{X}$ became a three-dimensional matrix with the shape of $(Width/8) \times (Height/8) \times 64$, which is represented by $\boldsymbol{M_{block}}$. $\boldsymbol{M_{DCT}}$ is a 64×64 matrix. By post-multiplying each column vector of $\boldsymbol{M_{DCT}}$ by $\boldsymbol{M_{block}}$, the element with the same frequency of all the blocks can be obtained. Let $\boldsymbol{Col_n}$ represent the nth column of the $\boldsymbol{M_{DCT}}$, and $\boldsymbol{Row_m}$ is the mth small block in the $\boldsymbol{M_{block}}$ $(m, n \in [1, 2, \cdots, 64])$. The process of right-multiplying $\boldsymbol{M_{block}}$ with $\boldsymbol{M_{DCT}}$ is shown in Eq. 6, which is the same as the process of the convolution operation. The window sliding of the convolution process ensures that each block in the $\boldsymbol{M_{block}}$ is traversed.

$$
\boldsymbol{M_{block}} \times \boldsymbol{M_{DCT}}
$$

$$
= \begin{bmatrix} \boldsymbol{Row_1} \times \boldsymbol{Col_1} & & \boldsymbol{Row_1} \times \boldsymbol{Col_{64}} \\ & \ddots & \\ \boldsymbol{Row_{64}} \times \boldsymbol{Col_1} & & \boldsymbol{Row_{64}} \times \boldsymbol{Col_{64}} \end{bmatrix}
$$

$$
= \begin{bmatrix} \sum_{i=1}^{64} Row_{1i} \times Col_{1i} & & \sum_{i=1}^{64} Row_{1i} \times Col_{64i} \\ & \ddots & \\ \sum_{i=1}^{64} Row_{64i} \times Col_{1i} & & \sum_{i=1}^{64} Row_{64i} \times Col_{64i} \end{bmatrix} \tag{6}
$$

In Eq. 6, Row_{mi} and Col_{ni} represent the ith element in Row and Col, respectively. This means that if each column of $\boldsymbol{M_{DCT}}$ is used to initialize the weight of an 8×8 convolution kernel respectively, and the stride of convolution is set to 8, then $\boldsymbol{X_{DCT}}$ can be obtained through convolution. ConvDCT is used to represent the convolutional layer above, so $\boldsymbol{X_{DCT}}$ could be computed by Eq. 7.

$$
\boldsymbol{X_{DCT}} = SDCT(\boldsymbol{X}) = ConvDCT(\boldsymbol{X}) \tag{7}
$$

Similarly, the convolutional layer ConvIDCT can be used to achieve IDCT of small blocks. $\boldsymbol{X_{IDCT}}$ is used to represent the result of IDCT of blocks, and Eq. 8 is gotten.

$$
\boldsymbol{X_{IDCT}} = SIDCT(\boldsymbol{X}) = ConvIDCT(\boldsymbol{X}) \tag{8}
$$

Since DCT and IDCT are inverse transforms to each other, ConvDCT and ConvIDCT are also inverses into each other. In other words, they satisfy Eq. 9.

$$
\boldsymbol{X} = ConvIDCT(ConvDCT(\boldsymbol{X})) \tag{9}
$$

Experiments show that the high forensics precision of adversarial examples can be achieved without transforming back to the spatial domain behind the Remainder layer. Therefore, in the proposed method of this paper, we only use the ConvDCT layer.

The weight of the convolution kernels of the ConvDCT layer will be updated. This means that as the training progresses, the input example X will gradually deviate from the DCT domain, and it will be transferred to a DCT-like domain that is more conducive to forensics. Besides, each convolution kernel also contains a trainable bias.

2.3   Remainder Layer

In the experiment of using JPEG compression error to detect adversarial examples, we find that the error generated by JPEG compression with a quality factor of 100 is the most effective. This means that each element of M_Q in Eq. 4 is 1. However, M_Q of JPEG compression has a fixed numerical range and is not specifically customized for forensics. In other words, M_Q with each element of 1 is not the best choice and it needs to be optimized with the training of the network.

It is noteworthy that the transformation domain images obtained with the ConvDCT layer include 64 feature graphs $(Width/8) \times (Height/8)$, each of which is composed of the same frequency parts of all the 8×8 blocks. In other words, when example X is feed to the ConvDCT layer, a matrix with the shape of $64 \times 8 \times 8$ will be gotten. A matrix with a shape of $64 \times 8 \times 8$ is initialized to 1, and it is registered as an updatable parameter. Each 8×8 submatrix is given a bias, then the Remainder layer is constructed.

2.4   Color Space Transformation

For color images, JPEG compression is performed in YCbCr space. Experiments show that forensics in the YCbCr color space is also better than RGB space. Let X_{RBG} denote samples in the RBG color space, and X_{YCbCr} to represent samples in the YCbCr color space. The process of color space transformation is very simple and can be expressed by Eq. 10.

$$X_{YCbCr} = \begin{bmatrix} 0.299 & 0.587 & 0.114 \\ -0.169 & -0.331 & 0.500 \\ 0.500 & -0.419 & -0.081 \end{bmatrix} \cdot X_{RGB} \\ + [\, 0 \quad 128 \quad 128 \,]^{\mathrm{T}} \tag{10}$$

The Cg channel of X is represented by X_{Cg}. It can be calculated by Eq. 11.

$$X_{Cg} = [-0.362 \quad 0.500 \quad -0.138] \cdot X_{RGB} + 128 \tag{11}$$

The process of Eq. 10 and Eq. 11 can be simulated by four convolution kernels with a shape of $1 \times 1 \times 3$, and the bias can be simulated with the bias of the convolution kernel to achieve good results. This layer is called the color space conversion layer and is expressed by the ConvTran Layer, then Eq. 12 is gotten.

$$X_{YCbCr} = ConvTran(X_{RGB}) \tag{12}$$

Similar to the ConvDCT and ConvIDCT layers, as the training process progresses, the weight value of ConvTran will gradually deviate from the value in Eq. 10. At the same time, X_{YCbCr} will no longer be an image of YCbCr color space strictly, but gradually enter another color space which is more suitable for forensics.

2.5 Feature Statistical Layer

The emergence of the global pooling layer [12] has brought many benefits to the neural network, such as reducing the number of network parameters, speeding up the training speed, and reducing the phenomenon of overfitting. The global pooling layer is usually divided into a global maximum pooling layer and a global average pooling layer, and their role is usually understood as downsampling. From the perspective of forensics, the maximum and average of the feature maps are statistical features. In this paper, the global pooling layer is expanded to obtain the statistical values such as variance, minimum, skewness, and kurtosis of the feature map. Then a new layer—Feature Statistical Layer—is gotten. The operation of the feature statistical layer on the data is shown in Fig. 2.

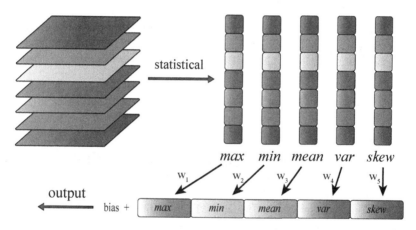

Fig. 2. Feature statistical layer. The colored parallelograms on the left represent the feature maps output by the convolution layer, and the colored squares on the right represent the feature values calculated on these feature maps. The colored rectangle at the bottom represents the one-dimensional vector formed by multiplying the features by the corresponding weights. (Color figure online)

In Fig. 2, $w_1 - w_5$ represents the weights of maximum, minimum, mean, variance, and skewness respectively. In the experiment, the validity of these five features has been verified. And bias is a bias term. After obtaining the statistical values of the feature maps, they are multiplied by the corresponding weights according to the feature type. Then, these features will be flattened into a one-dimensional vector. Finally, a bias is added to all of the features to obtain the final output. Out_{fs} is the output of the feature statistical layer. $max()$, $min()$, $mean()$, $var()$, and $skew()$ are used to calculate the maximum, minimum, mean, variance, and skewness respectively.

2.6 Group BN Strategy

The ConvDCT layer outputs 64 feature maps with the shape of $(Width/8) \times (Height/8)$, corresponding to 64 different frequencies. However, the numerical ranges of different frequencies are quite different. This difference spans several orders of magnitude, and it severely hinders the increase of the forensics accuracy.

A similar problem also appears in the feature statistical layer. After acquiring various types of feature values, there are also huge differences in the numerical ranges of different features. Relying only on their corresponding weights to balance this difference will be a slow process, and may even lead to non-convergence of the network, ultimately wasting a lot of computing resources.

These two differences are both generated between different types. To solve this kind of problem, we adopt the strategy of grouping Batch Normalization (BN [9]) layers. In the ConvDCT layer, the feature maps correspond to a set of 64 BN layers divided by frequencies. In the feature statistical layer, we adopt five independent BN layers corresponding to five types of features: maximum, minimum, mean, variance, and skewness. This solution is called a Group BN Strategy. Both the ConvDCT layer and the feature statistical layer mentioned above contain an additional Group BN Strategy.

3 Experimental Results

3.1 Experiment Introduction

The data set used in this paper is based on the ImageNet2012 data set [17]. First, we randomly selected 40,000 images from the ImageNet2012 dataset. 25,000 of these images were put into the training set, 5,000 into the validation set, and 10,000 into the test set. Next, we used FGSM [6], Deepfool [14], and C&W methods [2] to generate adversarial examples for these images on VGG16 [18], respectively. FGSM includes versions with disturbance coefficients of 2, 4, 6 and 8. In this way, a total of 6 child data sets are generated, each of which contains 50,000 examples for training, 10,000 examples for validation and 20,000 examples for the test. All experiments in this paper run on an Nvidia RTX 2080Ti Graphics Processing Unit (GPU).

3.2 Performance Evaluation

Model Efficiency. As mentioned above, the proposed method is a lightweight forensics model. This is not only because the training time for one epoch of the model is short, but also because the number of epochs required for training is small. Usually, one epoch of training is able to achieve the desired results. The accuracy and loss curves during the training on the C&W data set are shown in Fig. 3.

It can be seen from Fig. 3 (b) that the over-fitting phenomenon has occurred after the eighth epoch of the training process. At the same time, Fig. 3 (a) shows that the accuracy has also reached its maximum in the eighth epoch. At this time, the accuracy on the validation set is 95.18%. The learning rate of all the training processes in Fig. 3 is 0.0001. In order to achieve high accuracy with only one training epoch, we made a slight adjustment to the learning rate. The learning rate on the C&W dataset is set to 0.00095. The accuracy of each data set is shown in Table 1.

Table 1. Accuracies on Different Data Set

Attack method	FGSM($\epsilon = 2$)	FGSM($\epsilon = 4$)	Deepfool	C&W
acc(1)	99.16%	99.66%	98.54%	93.46%
acc(max)	99.64%	99.93%	99.31%	95.02%

In Table 1, acc(1) is the accuracy of only one period of training, and acc(max) represents the maximum accuracy achieved. By comparing the data of acc(1) and acc(max), it can be found that there is a gap of less than 2% between the accuracy achieved by only one epoch training and the maximum accuracy achieved by multi-epoch training. This is an acceptable range. Therefore, in practical application, the model proposed in this paper only needs one training epoch.

Module Effectiveness

Effectiveness of Group BN Strategy. The group BN strategy is necessary for the ConvDCT layer. The network cannot converge without the group BN strategy to normalize data in the DCT-like domain. After a long time of small fluctuations, the accuracy is still around 50%.

In the feature statistical layer, group BN strategy are not necessary for most features that have been adopted. Such features include maximum, mean, minimum, and variance. For some features, however, it is still necessary. Such features include skewness and kurtosis. In other words, if the feature statistical layer only includes maximum, mean, minimum and variance, the network can converge normally without the group BN strategy. Figure 4 shows the network training situation when these four feature values are used and the group BN strategy is not included, the four features and the grouping regularization strategy are used, and the network is trained with the skewness and the group BN strategy. These experiments are performed on the Deepfool data set.

In Fig. 4, comparing the accuracy curves of Fig.(a) and Fig.(b), it can be found that the group BN strategy not only speeds up the training process but

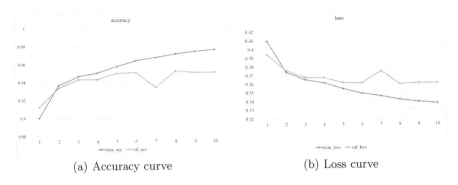

(a) Accuracy curve   (b) Loss curve

Fig. 3. The training process of the proposed model on the C&W dataset. The train_acc and val_acc are the accuracies of the model on the training set and validation set respectively, and the train_loss and val_loss are the value of the loss function on the training set and validation set respectively.

also helps to improve the accuracy. Comparing Fig.(b) and Fig.(c), it can be seen that skewness is very effective as an independent feature. It plays an important role to the final accuracy of the network. And it must be used together with the group BN strategy, so the group BN strategy in the feature statistical layer is also very important.

Effectiveness of ConvTran and ConvDCT. The comparison of the accuracy of the network without the ConvTran layer, the network without the ConvDCT layer, and the network without both is shown in Table 2. Here, only the FGSM data set with a disturbance coefficient of 2 is adopted.

In Table 2, FGSM(1) represents the accuracy after one epoch of training on the FGSM data set, and FGSM(max) is the highest accuracy achieved on the FGSM data set during 10 epochs. The following Deepfool(1), Deepfool(max), C&W(1), and C&W(max) have similar meanings. Comparing the first and fourth rows, the YCbCr color space is very important for the ConvDCT layer. It is as if the YCbCr color space exists specifically for the DCT domain. The data in the second and third lines indicate that ConvDCT layer is one of the basic structures of the model proposed in this paper, and it plays the most important role in the performance of forensic. Without its presence, the neural network is even completely incapable of detection in some data sets, such as the C&W data set.

Effectiveness of Feature Statistical Layer. The effectiveness of the feature statistical layer is directly related to the selected features. When we select only one of the maximum or mean, it is simply a global pooling layer with weights and bias. In the case that different features are selected by the feature statistical layer, the accuracy of the network is listed in Table 3. For the convenience of presentation, Table 3 contains the results of only one training epoch.

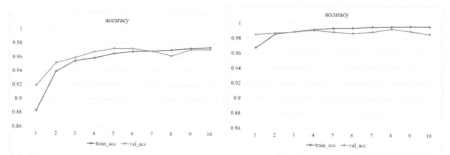

(a) Maximum, mean, minimum, and vari-   (b) Maximum, mean, minimum, and variance, without group BN strategy   ance, with group BN strategy

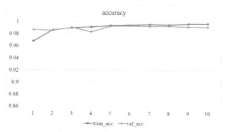

(c) Maximum, mean, minimum, variance,
skewness, with group BN strategy

Fig. 4. The accuracy curves of Grouping BN strategy of the feature statistical layer. The train_acc and val_acc are the accuracies on the training set and validation set respectively.

For the comparison with traditional modules, the first and second rows in Table 3 do not include parameters. In other words, the first row uses the global maximum pooling layer and the second row uses the global average pooling layer. Comparing with the following lines, the first two lines have poor forensic effect. This shows that the feature statistical layer is superior to the global pooling layer, and the forensics ability of a single feature is limited for adversarial examples. The following rows of data prove the effectiveness of the combined features. The data in the last row is not significantly improved compared with the second-to-last row, indicating that the network performance has reached the bottleneck after the number of features reaches five. It is reasonable to select five features.

3.3   Experiment of Comparison

On the datasets generated by FGSM ($\epsilon = 2, 4, 6, 8$), Deepfool, and C&W method, we compare the detection rates of the proposed method, Feature Squeezing [23], and ESRM [13]. Their accuracies are shown in Table 4.

In Table 4, on FGSM ($\epsilon = 2, 4, 8$), Deepfool, and C&W data sets, the accuracies of the proposed method only trained for one epoch exceeded the final

Table 2. Effectiveness of ConvTran layer and ConvDCT layer

Attacke method	FGSM(1)	FGSM(max)	Deepfool(1)	Deepfool(max)	C&W(1)	C&W(max)
No ConvTran	96.26%	96.88%	89.44%	91.38%	74.71%	77.53%
No ConvDCT	70.59%	74.67%	55.60%	58.25%	50.58%	51.30%
No Both	60.32%	65.86%	58.88%	60.47%	50.43%	50.93%
Complete	99.16%	99.64%	98.54%	99.07%	93.46%	95.02%

Table 3. Contrast of Different Features

Attack method	FGSM($\epsilon = 2$)	FGSM($\epsilon = 4$)	FGSM($\epsilon = 6$)	FGSM($\epsilon = 8$)	Deepfool	C&W
max	96.52%	98.19%	98.96%	99.63%	90.15%	82.38%
mean	96.06%	98.47%	94.61%	99.76%	89.62%	70.60%
max, mean, min	**99.34%**	99.81%	99.89%	99.93%	98.49%	89.54%
max, mean, var	99.03%	**99.89%**	99.86%	99.94%	98.60%	93.42%
max, mean, min, var	99.26%	99.62%	99.84%	99.94%	**98.62%**	91.53%
max, mean, var, skew	99.24%	99.82%	99.85%	99.96%	98.05%	92.97%
all	99.16% ·	99.66%	99.03%	**99.98%**	98.54%	**93.46%**
all, kurt	99.16%	99.87%	**99.92%**	**99.98%**	98.12%	93.27%

accuracies of Feature Squeezing and ESRM. On the C&W data set, the highest accuracy of the proposed method is still greatly improved compared with the accuracy of the first epoch.

Feature squeezing is essentially an example repair method. Its approach is to try to erase the adversarial of examples by some traditional signal processing methods so that the original classifier can recognize correctly. Of course, the different recognitions of the original classifier can be used to detect the adversarial examples. However, this method makes the detection accuracy directly related to the amplitude of adversarial perturbations. As shown in Table 4, it performs well on the C&W datasets. With the increase of perturbation coefficient ϵ on the FGSM dataset, its accuracy shows a significant downward trend. We calculated the adversarial distortions of each dataset, which is shown in Table 5.

The adversarial distortion [20] is a measure of the perturbations of the adversarial examples compared with its corresponding natural examples. Let X_{ad} represent the adversarial example, and X_{na} represents the natural example corresponding X_{ad}. n is the number of pixels in sample X. The distortion X_{Dis} can be calculated by Eq. 13. The larger the distortion is, the more adversarial noise is added. The distortions in Table 5 are consistent with the detection accuracies of Feature Squeezing on these data set.

$$X_{Dis} = \|X_{ad} - X_{na}\|_2/\sqrt{n} \tag{13}$$

The ROC curves of the proposed method (training for one epoch), ESRM, cascade classifier [11], and feature squeezing [23] on the Deepfool dataset are shown in Fig. 5. The accuracy of the cascade classifier is 79.54%. According to Table 5 and Fig. 5, except for the method proposed in this paper, ESRM has the best detection performance on each data set.

Table 4. Comparison of Accuracies of Different Methods

Attack method	Detection method	Detection accuracy
FGSM($\epsilon = 2$)	The Proposed Method(1)	99.16%
	The Proposed Method(max)	**99.64%**
	Feature Squeezing [23]	67.45%
	ESRM [13]	98.10%
FGSM($\epsilon = 4$)	The Proposed Method(1)	**99.66%**
	The Proposed Method(max)	-
	Feature Squeezing [23]	64.64%
	ESRM [13]	98.53%
FGSM($\epsilon = 6$)	The Proposed Method(1)	**99.03%**
	The Proposed Method(max)	-
	Feature Squeezing [23]	57.60%
	ESRM [13]	**99.03%**
FGSM($\epsilon = 8$)	The Proposed Method(1)	**99.98%**
	The Proposed Method(max)	-
	Feature Squeezing [23]	55.85%
	ESRM [13]	99.35%
Deepfool	The Proposed Method(1)	98.54%
	The Proposed Method(max)	**99.07%**
	Feature Squeezing [23]	61.12%
	ESRM [13]	95.13%
C&W	The Proposed Method(1)	93.46%
	The Proposed Method(max)	**95.02%**
	Feature Squeezing [23]	92.09%
	ESRM [13]	92.87%

ESRM is a traditional detection scheme, which is based on the traditional steganalysis method. And is very sensitive to the adversarial distortions. On the FGSM data set, with the increase of perturbation coefficient ϵ, its detection performance is significantly improved. At the same time, it does not perform as well on the Deepfool and C&W datasets. Due to the separation of the feature extraction process and training process, this detection scheme cannot be supported by GPU parallel computing acceleration. With high time complexity, the training process is very long on large data sets. In contrast, the proposed method needs only one training epoch to achieve the desired effect of the application. On the data set including 50,000 examples used in this paper, this process takes only no more than 5 min.

The proposed method uses a powerful DCT-like domain and the feature statistical layer to achieve high-precision forensics of adversarial examples in a short time. On the whole, compared with the existing forensics methods, the proposed program has obvious advantages in training time and forensics effect.

Table 5. Distortions of Different Dataset

Attack method	Distortion
FGSM($\epsilon = 2$)	1.99
Deepfool	1.11
C&W	0.86

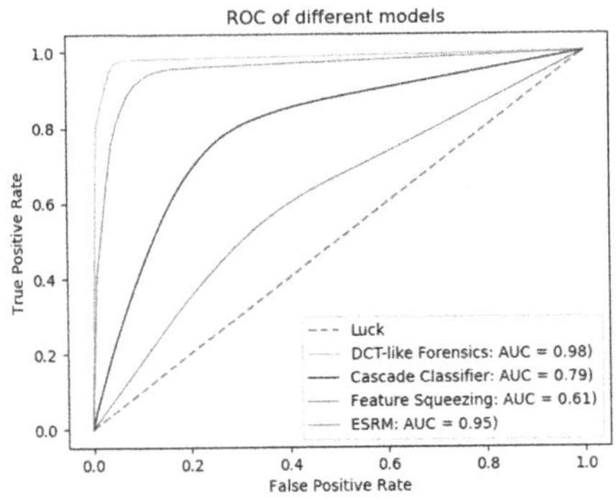

Fig. 5. ROC Curves of Different Methods

4   Conclusion

In this paper, a lightweight forensics model of adversarial examples based on the DCT-like domain and the feature statistical layer is proposed. It cleverly combines the quantization error in JPEG compression, manual feature extraction and deep neural network. Inspired by the quantization error in JPEG compression, we designed the color space transformation layer and the DCT-like layer based on the convolutional layer. The feature statistical layer is used to effectively obtain the feature value in the frequency domain. The experiment demonstrates the superiority of the proposed model in the forensics of adversarial examples. In future work, we will explore more features and functions of the DCT-like domain.

References

1. Bochkovskiy, A., Wang, C.Y., Liao, H.Y.M.: Yolov4: optimal speed and accuracy of object detection. arXiv preprint arXiv:2004.10934 (2020)
2. Carlini, N., Wagner, D.: Towards evaluating the robustness of neural networks. In: 2017 IEEE Symposium on Security and Privacy (SP), pp. 39–57. IEEE (2017)
3. Carrara, F., Becarelli, R., Caldelli, R., Falchi, F., Amato, G.: Adversarial examples detection in features distance spaces. In: Leal-Taixé, L., Roth, S. (eds.) ECCV 2018. LNCS, vol. 11130, pp. 313–327. Springer, Cham (2019). https://doi.org/10.1007/978-3-030-11012-3_26
4. Fridrich, J., Kodovsky, J.: Rich models for steganalysis of digital images. IEEE Trans. Inf. Forensics Secur. **7**(3), 868–882 (2012)

5. Geng, Q., Zhang, H., Qi, X., Yang, R., Zhou, Z., Huang, G.: Gated path selection network for semantic segmentation. arXiv preprint arXiv:2001.06819 (2020)
6. Goodfellow, I.J., Shlens, J., Szegedy, C.: Explaining and harnessing adversarial examples. arXiv preprint arXiv:1412.6572 (2014)
7. Grosse, K., Manoharan, P., Papernot, N., Backes, M., McDaniel, P.: On the (statistical) detection of adversarial examples. arXiv preprint arXiv:1702.06280 (2017)
8. Huang, G., Liu, Z., Van Der Maaten, L., Weinberger, K.Q.: Densely connected convolutional networks. In: Proceedings of the IEEE Conference on Computer Vision and Pattern Recognition, pp. 4700–4708 (2017)
9. Ioffe, S., Szegedy, C.: Batch normalization: accelerating deep network training by reducing internal covariate shift. arXiv preprint arXiv:1502.03167 (2015)
10. Kurakin, A., Goodfellow, I., Bengio, S.: Adversarial examples in the physical world. arXiv preprint arXiv:1607.02533 (2016)
11. Li, X., Li, F.: Adversarial examples detection in deep networks with convolutional filter statistics. In: Proceedings of the IEEE International Conference on Computer Vision, pp. 5764–5772 (2017)
12. Lin, M., Chen, Q., Yan, S.: Network in network. arXiv preprint arXiv:1312.4400 (2013)
13. Liu, J., et al.: Detection based defense against adversarial examples from the steganalysis point of view. In: Proceedings of the IEEE Conference on Computer Vision and Pattern Recognition, pp. 4825–4834 (2019)
14. Moosavi-Dezfooli, S.M., Fawzi, A., Frossard, P.: DeepFool: a simple and accurate method to fool deep neural networks. In: Proceedings of the IEEE Conference on Computer Vision and Pattern Recognition, pp. 2574–2582 (2016)
15. Mopuri, K.R., Ganeshan, A., Babu, R.V.: Generalizable data-free objective for crafting universal adversarial perturbations. IEEE Trans. Pattern Anal. Mach. Intell. 41(10), 2452–2465 (2018)
16. Peng, F., Li, J., Long, M.: Identification of natural images and computer-generated graphics based on statistical and textural features. J. Forensic Sci. 60(2), 435–443 (2015)
17. Russakovsky, O., et al.: ImageNet large scale visual recognition challenge. Int. J. Comput. Vis. 115(3), 211–252 (2015)
18. Simonyan, K., Zisserman, A.: Very deep convolutional networks for large-scale image recognition. arXiv preprint arXiv:1409.1556 (2014)
19. Sundermeyer, M., Schlüter, R., Ney, H.: LSTM neural networks for language modeling. In: Thirteenth Annual Conference of the International Speech Communication Association (2012)
20. Szegedy, C., et al.: Intriguing properties of neural networks. arXiv preprint arXiv:1312.6199 (2013)
21. Wang, J., Li, T., Luo, X., Shi, Y.Q., Jha, S.K.: Identifying computer generated images based on quaternion central moments in color quaternion wavelet domain. IEEE Trans. Circuits Syst. Video Technol. 29(9), 2775–2785 (2018)
22. Wang, J., Wanga, H., Li, J., Luo, X., Shi, Y.Q., Jha, S.K.: Detecting double jpeg compressed color images with the same quantization matrix in spherical coordinates. IEEE Trans. Circuits Syst. Video Technol. 30, 2736–2749 (2019)
23. Xu, W., Evans, D., Qi, Y.: Feature squeezing: detecting adversarial examples in deep neural networks. arXiv preprint arXiv:1704.01155 (2017)

Author Index

Printed in the United States
By Bookmasters